General Systems Thinking
Its Scope and Applicability

THE NORTH HOLLAND SERIES IN
General Systems Research
Dr. George Klir, *Editor*

1 **Rosen**   Fundamentals of Measurement
                and Representation of Natural Systems

2 **Varela**   Principles of Biological Autonomy

3 **Zeleny**   Autopoiesis:
                A Theory of Living Organization

4 **Bowler**   General Systems Thinking:
                Its Scope and Applicability

General Systems Thinking

Its Scope and Applicability

Series Volume 4

T. Downing Bowler
Bradford College
Bradford, Massachusetts

North Holland
New York • Oxford

Elsevier North Holland, Inc.
52 Vanderbilt Avenue, New York, New York 10017

Sole Distributors outside the USA and Canada:
Elsevier Science Publishers B.V.
P.O. Box 211, 1000 AE Amsterdam, The Netherlands

© 1981 by Elsevier North Holland, Inc.

Library of Congress Cataloging in Publication Data

Bowler, Downing T
    General systems thinking: its scope and applicability.
    (The North Holland series in general systems research; 4)

    Bibliography: p.
    Includes index.
    1. System theory. I. Title.
Q295.B68   003   80-20469
ISBN 0-444-00420-3

Desk Editor John Haber
Design Series
Art Editor Glen Burris
Production Manager Joanne Jay
Compositor Intergraphic Technology, Inc.
Printer Haddon Craftsmen

Manufactured in the United States of America

*Dedicated to
Janet F. Bowler*

Contents

Preface		**xi**
Chapter 1	**General Characteristics of Systems**	**1**
	General Characteristics	2
Chapter 2	**Simple Stable Order Systems**	**10**
	Atomic Systems	12
	Molecular Systems	19
Chapter 3	**The Universe as a Physical System**	**26**
	Continuum of Differentiation	29
	Continuum of Uniqueness	30
	Closed and Open Systems	30
	Description of a Specific System	31
Chapter 4	**Steady State (Morphostatic) Systems**	**33**
	Equilibration	33
	Competition, Cooperation, and Control	36
	System Controls	38
	Specialization	41
	Description of a Specific System	44
Chapter 5	**The Universe as a Biosphere**	**45**
	The Earth as an Open System	46
	Subsystem Autonomy and System Dominance	49
	Levels of Systems	52
	Competition and Cooperation	52
	Selectivity and Discrimination	53
	Control Systems	55

Chapter 6	**Psychic Systems**	**63**
	Behavioral Psychology	67
	Connections and Relations	72
	Psyche and System	73
	Personality and System	83
Chapter 7	**General Systems Theory and Psychic Systems**	**89**
	Developmental Continuum	92
	Value Judgments	96
	The Nature of Subjective Experience	97
	Methodology	109
Chapter 8	**Equilibration, Stress, Development, Inertia, and Defense**	**114**
	Learning and Central Control	115
	Complex Learning Environment	117
	Motivation for Integration	118
	Limits of Integration	119
	Parallel Development in Psychic and Social Systems	120
	Basic Needs in Understanding Psychic Organization	121
	Demonization and Development	124
Chapter 9	**Psychosocial Interaction, Personal Identity, and Personal Meaning**	**127**
	Diametric Polarities	128
	Relations and Identity	130
	Problem of Self-Motivating Development	131
	Personal Identity	133
	Identification	134
	Recognition	136
	Loss of Identity	138
	Self-Consciousness	139
	Subsystem Autonomy and System Dominance	140
	Sense of Identity	142
	Personal Meaning	143
	Transcendent Perspective	145
Chapter 10	**Microsocial Systems**	**148**
	Bonding	150
	A General Theory of Human Systems	150
	Differentiation	153
	Integration	158
	Problem Perception	161
	Specialization and Organization	162
	Socialization	167
	Organization	171

Chapter 11	**Macrosocial Systems**	**176**
	Some Central Problems	177
	Primitive (Simple Macro-) Social Systems	178
	System Autonomy	181
	Subsystem Autonomy	183
	Sources of Variety	183
	Transition from Simple to Complex	188
	Necessary Sequence, Not Necessary Development	191
	Characteristics of Civilizations	191
	Stages of Development	197
	Some Applications of General Systems	202
	Process of Development	204
	Varieties of Equilibration	206
	Residual Maps	208
	Degrees of Boundary and Degrees of Existence	208
	World Social System	210
Chapter 12	**General Systems Theory as Philosophy**	**211**
	Mythopoeic World Views	212
	Rational World Views	212
	The Fragmentation of Knowledge	214
	Civilizational Disintegration	215
	The Function of Generalizations	216
	Relativism	217
	Relational Universals	218
	Mathematics: The Pure Science of Possible Relations	218
	General Systems Theory: An Existential Science of Possible Relations	218
	Some Implications	222
	Some Problems	225
Index		**231**

Preface

This book attempts to organize the basic concepts that underlie general systems thinking and to demonstrate their universal scope and applicability. It represents 20 years of teaching, research, and writing in search of an overview consistent with the present state of scientific knowledge. The primary motivation for this work derives from two convictions. First, there is a need for a comprehensive perspective on human beings and their world that would make our fragmented information more readily available for the intelligent conduct of life. Second, common patterns have emerged in the information provided by the specialized sciences that make such an overview possible.

The approach focuses on the concept of system as organized process. The word "system" means literally to cause to stand together. Throughout the book, various levels of entities are examined in terms of the organization, processes, relations, dynamic tensions, and so on that cause them to stand together as recognizable entities. At each level, systems (entities) are analyzed with the concepts provided by the relevant fields of specializaion and correlated with the concepts of general systems thinking. The perspective that emerges shifts from an object-oriented to a process-oriented model. The traditional quest for the solid objects of reality is replaced by an understanding of an entity as a concentration of organized processes that has a relative degree of autonomy and functions as a subsystem in some number of higher-order systems. As we shall see in Chapter 12, this perspective has been emerging through most of the present century in one form or another. In all its various forms, it represents the need of the social system to equilibrate the proliferation of fragmented specializations with suitable integrative processes.

The first three chapters focus on information concerning the nature of the physical universe and correlate that information with some of the basic systems generalizations. Chapter 1 concentrates on electromagnetic energy and subatomic particles, and Chapter 2 carries the application of these basic concepts to the levels of atoms, molecules, and macromolecules. Chapter 3 applies the same kind of analysis to the universe as a physical system. Chapters 4 and 5 focus on biosystems and the universe as a biosphere, respectively. Chapters 6–8 correlate systems concepts with information from various areas of psychological investigation under the general heading of psychic systems. (The reasons for the use of this term are explained in Chapter 6.) Chapter 9 is a transition from psychic to social systems and emphasizes the intimate relation between these two levels of systems. Chapter 10 moves up to the level of smaller (micro-) social systems, and Chapter 11 examines whole societies as macrosocial systems. Finally, Chapter 12 presents the case for general systems theory as a philosophy of world and life, includes a summary of the systems concepts used thoughout the book, and sets forth some of their implications for selected problems.

Much of the material in the final chapter was presented in a paper to the August 1979 meeting of the Society for General Systems Research and has been adapted with their permission. I would like to acknowledge my indebtedness to the faculty of Bradford College, past and present, who made my progress through so many diverse fields of knowledge manageable.

General
Systems
Thinking
Its Scope and Applicability

Chapter 1

General Characteristics of Systems

General systems theory is, in simple terms, an attempt to develop useful generalizations about systems. It is based on the assumption that all systems have some characteristics in common and that those characteristics which are not common to all systems provide the terms for differentiating among types of systems. All generalizations are concerned with classes of entities and all words, except proper nouns, are generalizations. "Chair" is such a word, and refers to a class of objects that share sufficient characteristics to be identified as chairs. However, there are many different kinds, or subclasses, of chairs, for example, overstuffed, straight-backed, or folding. Each subclass includes chairs with sufficient characteristics in common to distinguish it from other subclasses, and yet these subclasses too have subclasses. For example, there are many kinds of folding chairs, such as lawn chairs and the wooden kind used in meeting halls. However, if we work in the opposite direction, we can include all chairs in a larger generalization, such as furniture, all furniture in the class of artifacts, or man-made objects, and so on.

The word "system" is another generalization refering to a class of entities. What we need to know is the kind or kinds of entities designated by the term. At this time, there is no universal agreement concerning the proper referents of this word, but for the purposes of this book we shall use it with its most general signification.

The word "system" is derived from two Greek words meaning to cause to stand together or to place together. In this sense, system is the opposite of randomness or chaos. In most ancient mythologies, people expressed their intuitive awareness of the tension between the ordered universe, as they knew it, and the chaos that resulted when some of that organization broke down. It is common in such mythologies for the preservation of order to be attributed to a god or group of gods who were strong enough to maintain it against the forces of chaos. This is a belief still shared, that there is an ordered universe rather than chaos only because things are held together in sets of relationships. This intuitive awareness has been informed by an ever-increasing knowledge of the nature of these entities and their relations. Everything that has been analyzed consists of parts that are "caused to stand together" by sets of relations. When the parts are themselves analyzed, they, too, are seen to be systems of parts that "stand together" in certain sets of relations. In other words, everything that is recognizable as an entity turns out to be a system. However, does a word that signifies everything have any use? It might seem that any word or phrase that applies to everything could not give us useful information about anything. Yet there are words having as their referents those characteristics that are the common ground or unity of all things. Such words are simply larger generalizations than others and tell us something about the nature of existence. General systems theory is the highest level of generalization short of formal disciplines such as mathematics and logic. Nevertheless, the goals are the same as they are for all pursuits of knowledge; that is, we must delineate those characteristics common to the class and then identify subtypes by their unique characteristics.

General Characteristics

In order to find the greatest common denominator of all systems, it is necessary to begin with the simplest of all systems, and in order to do this, it is necessary to discuss briefly the nature of energy and of subatomic particles. Man is aware of *energy* in many forms, but the most fundamental is electromagnetic radiation, which includes not only sunlight, but much more than what we think of as light. What we humans experience as light is a small portion of the whole spectrum of electromagnetic energy, which is known and used. All forms of this energy travel or radiate at the same speed (the speed of light: 186,000 miles per second). The differences in manifestation are the result of variations in the wavelength and frequency of oscillation. Electromagnetic energy always travels at the same speed but may vary as to how far and how fast it oscillates from side to side. What is commonly spoken of as radiation, such as gamma and x rays, oscillates short distances and very rapidly penetrates almost anything, and is highly destructive to many systems. Visible light is made up of the colors of the spectrum (rainbow), each having a different wavelength. Violet oscillates the most rapidly and has the shortest wavelength, while red oscillates the most slowly and has the longest wavelength of the visible colors. All colors have long wavelengths and oscillate slowly relative to the dangerous forms, radiation, and hence visible light does not pass through most solid objects. Instead the objects cast shadows.

There are other forms of this basic energy having much longer wavelengths that oscillate much more slowly. These are used for radio and television signals and as electrical energy in wired circuits. The entire spectrum is often represented in diagrammatic form, as in Figure 1. This energy, in its various forms, appears more and more likely to be the ultimate ground of all natural forms of existence of which man has any knowledge or experience, and there is much about it that we do not and may never know. This is not to suggest that the universe was ever only energy and then became matter. Electromagnetic energy is, however, the ultimate ground in the sense that all forms of existence can be broken down into more simple forms until one reaches the final product, the release of energy. Under the right circumstances, matter and energy are transformed from one to the other in accordance with Einstein's formula $E=mc^2$. (E represents energy in ergs, m represents mass in grams, and c represents the speed of light in centimeters per second. Because the speed of light is 30 billion centimeters per second, even a small amount of mass would produce a very large amount of energy. We can thus consider mass or matter as "packaged energy." Just how it is "packaged" is not known. It is enough for our purposes to know that elecromagnetic energy can be transformed into particles and that a large particle is some kind of system of smaller particles. That is, when heavy particles break down to lighter particles, they emit energy (photons) and small particles. In other words, even at the level of subatomic particles, systems appear to have two fundamental characteristics: energy and relations. A *relation* is simply a connection between two or more systems that operates as a constraint on the behavior of the system involved. *Constraint* must be understood against the background of possible variety. Individuals are capable of a wide variety of behavior, but the more relations they are part of, the more constraints limit their behavior.

Within the universe as we know it, to be is to be related. Entities or systems exist only insofar as they are related to something else. An entity that is related to nothing cannot be said to exist, because there is no way to describe its presence in any situation. One might argue that an entity can exist

Figure 1
The electromagnetic spectrum.

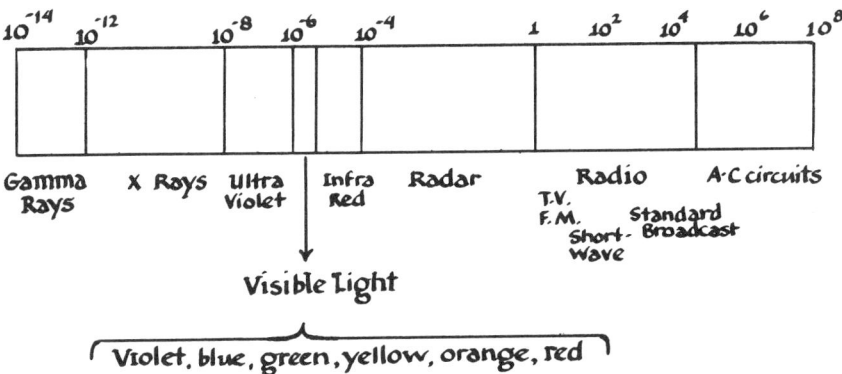

even if we do not perceive it, but such an entity would make no difference to anything else. In terms of the ordered universe, it would be a nonentity.

This becomes more clear if we move up to a higher level of complexity about which more is known. The nucleus of an atom is made up of protons and neutrons (heavy particles) bound together by nuclear force. If we regard these particles as systems of "packaged" energy, then we can say that the nucleus is a more complex system of energy, particles, and their relations. It is assumed that energy is somehow constrained in the formation of the particles, and we know that the particles are constrained by nuclear force in the formation of the nucleus.

It is necessary to understand relations as constraints eliminating other possibilities, which always exist. For example, protons all have positive charges and therefore repel each other, but that behavior is constrained by the relation of nuclear force that binds them together. In general, we can say that systems are constituted of the interrelations of their subsystems and are dependent on forms of energy.

All systems appear to contain opposing forces in some degree of balance. We can assume that there is some degree of equilibration between the electromagnetic repulsion among protons and the nuclear force that binds them together, but so little is known about the inner structure of the nucleus that it will be better to consider the atom as a whole. It is then easier to see that the electromagnetic attraction between the positively charged protons and the negatively charged electrons would result in their being drawn together if it were not balanced by the kinetic energy of the electrons.

The same balancing of forces is used in space travel. To put a spaceship into orbit, enough kinetic energy must be generated to balance the gravitational pull between the ship and the earth. For the spaceship to leave the earth and go to the moon, its kinetic energy must be increased to overcome the gravitational pull; that is, it must attain escape velocity. For a spaceship to orbit the moon, it must enter the moon's gravitational field with enough kinetic energy to neither whip off the other side nor drop to the surface of the moon. And for a ship to land on the moon, it must lose velocity (kinetic energy) and allow the gravitational pull to be the greater force.

In general, we can say that it is the balancing of opposing forces that makes systems possible. Such opposing forces, constraints, or processes will be referred to as *polarities*, for it is helpful to see them as the extremes of a continuum between which the system must be in some degree of relative balance. An exact balance is generally referred to as an equilibrium, but this is certainly not a universal condition for systems. However, a tendency toward the balancing of polar extremes does appear to be universal, and this shall be referred to as *equilibration*. Throughout the book, equilibration should be understood as referring to processes of balancing polarities and never to a state of equilibrium.

This writer is in agreement with those who reject the idea that equilibrium

General Characteristics

is an adequate model for the most complex systems, such as psyches and societies, but processes of equilibration are not the same as equilibria. Tightrope walkers are equilibrists although they never attain a state of equilibrium. They are constantly balancing, that is, equilibrating, even though their goal is not to arrive at a state of equilibrium. They come closest when the act is over and they step onto the stand at either end of the rope. We shall see when we turn to the most complex systems that they can never get "off the rope" without disintegrating. This, however, does not exclude them from the endless process of equilibrating their polar extremes. Motion invariably requires energy of some sort, so systems not constrained by some relations tend toward increasing randomness or entropy. Hence, in any system there must be some degree of equilibration in the acquisition, distribuion, use, and expenditure of energy.

So far, what has been indicated is that a system consists of forms of energy ordered by constraining relations that tend to equilibrate among their polar extremes, but this is not enough. One of the most basic polarities which runs through the whole gamut of systems is the tension between an *entity* and the *system* in which it participates. An entity taken utterly by itself tends to behave in certain ways. This is true even at the most simple level. According to the law of inertia, a body at rest tends to stay at rest, and a body in motion tends to stay in motion unless acted upon by some outward force. On the other hand, if an entity is to participate in a system, then it must become related, that is, constrained, within that system. This is variously referred to as a polarity between individuality and participation, self-assertion and integration, or subsystem autonomy and system domination. Whatever the terms, the fact of this universal polarity is something we shall encounter. It appears in Paul Tillich's philosophical theology as the fundamental self–world polarity of all existence. The law of inertia refers to an entity in an ideal vacuum but, in actuality, there are almost always some outward forces acting on an entity. That is, all entities are parts of some system even if that system is the whole universe. Of course, there are vast differences in the relative extent of subsystem autonomy and system dominance from system to system. A particle traveling through space may have long periods of time with little or no external interference, while a particle within the nucleus of an atom may have rather long periods of time with relatively little autonomy. Consequently, every system must be understood in terms of the relative balance (degree of equilibration) between these polar extremes.

By its very nature this polarity requires the concept of *boundary*. If an entity is recognizable as such, it must be describable in such a way as to differentiate it, in some degree, from other entities and from the system in which it participates. If there were no boundaries, there would be no differentiation, only random homogeneity. Interaction presupposes relatively differentiated entities capable of entering into relations of some sort. However, boundary must be understood as a matter of degree. There are no absolute boundaries. While the polarity between relative autonomy and system dominance requires the idea of boundary, it also requires that the boundary not be absolute. This leads us to distinguish among three sets of relations. There are the internal relations that constitute an entity as such, external relations among entities (e.g., stubbing one's toe on a rock or the reflection of light from a surface), and relations that cross the boundary and are usually some form of transmission of energy or matter. What is transmitted may be that which comes in (input), that which goes out (output), or something shared, such as electrons in covalent bonding among atoms. However, systems are not always boundary to boundary, and so we need the concept of interface, denoting an area between the boundaries of systems.

So far systems have been described in terms of energy and relations, but it is more common to discuss systems in terms of *structure*. However, at the present time, the word "structure" presents something of a problem. Structure is most often used to refer to the manner or form in which something has been constructed, such as bone structure, the structure of a house, or the structure of government. With the advent of structuralism, however, the word has received a new emphasis.

The model for this new emphasis comes from mathematics and is a quest for the abstract forms behind overt or specific forms. Whereas the human body could be described by a specific structure, one would describe the general determinations of transformation and development for all human bodies by a structure in the abstract sense. That is, the actual bodies may vary in many ways but always within the limits of a general form for human bodies. The importance of structuralism in this context is it considers entities as constructed (logically or physically) of existing entities and determined in changes from one level to another. In other words, systems tend to determine themselves in that they are governed by rules of transformation and development. In attempting to explain structuralism, Piaget defines it as a "system of transformation."[1] In this book the term "structure" will refer to the relatively enduring assemblies of relations within a system or, which is the same thing, among systems.

A structure can always be analyzed into a set of relations. A set of relations constitutes a structure when their systemic nature regulates the form, transformation, and/or development of a system. Further, because all systems are involved in some degree of motion and change, structure must always be seen in relation to *process*. The word "process" will be used to refer to the sequential states of a system and largely as synonymous with change, except that the former places more emphasis on the ordered nature of change. There is no system so rigid or static that it is not in some process at some level of its relations. An abandoned machine, for example, is in process of adjusting

---

[1] Piaget, Jean, 1970. *Structuralism.* New York: Basic Books.

to temperature changes and processes of disintegration. Atoms and molecules are in constant motion, and even mountains grow and wear away. Of course, there are different kinds of processes, such as those normal functions that maintain a system and those that change a system. However, the evidence seems to indicate that, in terms of the model of the universe as primarily energy and relatedness, structure must be seen as either the more or less stable moments within ongoing processes or the system of relations regulating or determining the processes.

Reference has already been made to variety in the context of constraints and relations. "Variety" is an important term for arriving at an understanding of systems because other relations and arrangements are always possible. If the constraints of a system break down, the parts of that system are then free to become involved in other patterns of behavior or relations. When a radioactive nucleus breaks down, for example, some of its parts are released as energy or particles. As long as the parts remain within the system, they are being constrained from other varieties of behavior. Any system is, therefore, *constrained variety*. A system is a set of specific relations and is always exclusive in relation to the variety of possibilities. In fact, any order, by its very nature, is exclusive and constrained from randomness. "Exclusive" means, in this context, selective of what is allowable and what is not allowable within the given order. In other words, any system or order discriminates by its very nature against some of the available possibilities.

However, just as there is no absolute boundary, there is no absolute exclusiveness. If systems were absolutely rigid, there would be no universe as we know it. Exclusiveness is also a matter of degree, and there is always some possibility for variation without destroying the system altogether. This may consist of such minor adjustments as changes of energy levels or temperature or modifications of internal relations due to changes in the environment. Other variations may come in the form of growth, that is, the addition of parts while maintaining the basic pattern of relations, or in the form of synthesis, in which two or more systems integrate to form a larger system. In the latter case, the integrating systems may retain much of their original form (relative autonomy) while suffering some modification as a result of new constraints on the larger system; the bonding of atoms is an example.

It is important to distinguish between growth and synthesis because the former is concerned with the elaboration or development of the form of an existing system and the latter with the creation of a new and larger system. Both may be creative processes and involve *novelty*. This is an important concept and one to which we will return again and again, but it must be understood in relative terms. Novelty, in this context, does not refer to something totally new in the universe. Rather, the system that is synthesized will have characteristics that are novel relative to those of the simple systems of which it is formed. This distinction is essential to understanding the relation of the levels of systems. For example, particles have characteristics that pure energy does not have, nuclei characteristics that particles do not have, atoms characteristics that nuclei do not have, and molecules characteristics that atoms do not have. New levels can emerge because of the novel characteristics that have emerged in the preceding stage. In general, systems cannot develop from subsystems to supersystems without the novel characteristics of the intervening systems. For example, particles do not become molecules without the novel characteristics of atoms. As we will see, this holds true even for the most complex systems. The old argument whether the whole is more than the sum of its parts can, within this context, be easily set aside. The whole has some characteristics and properties that are novel relative to the characteristics and properties of its parts.

There are three other characteristics so commonplace that they may well be overlooked or, if they are noticed in some systems, may be overlooked in others. The chemical laws dealing with gases are concerned primarily with these three characteristics: pressure, temperature, and volume. The universality of these characteristics may be more clear if they are translated into other terms. *Temperature* is a measure of the amount of energy and is directly effected by population density, among other things. *Pressure* is a measure of impact and is directly effected by both temperature and population density. *Population density* is a measure of systems per volume and is directly effected by changes in volume or numbers of systems. It may not be possible, least at this point, to reduce the interaction of other systems to precise mathematical formulas, but it will become clear as we go on that the interrelations of temperature, pressure, and population density are central to the understanding of a variety of systems.

One final characteristic that must be mentioned here is *hierarchic organization*. The word "hierarchy" will refer here to any set of relations in which units are organized into more inclusive units. For example, language is hierarchic because words are organized into sentences, sentences into paragraphs, paragraphs into chapters, and so on. As another example, particles are organized into atoms, atoms into molecules, molecules into macromolecules, and so forth. All systems are hierarchic in one way or another, and complex systems are hierarchic in a number of ways. All hierarchies are concerned with relations, but some are concerned more with structure, others with functional aspects of the system, such as information, control, energy, acquisition, and maintenance. More complex systems often exhibit overlapping hierarchies, so that some units are involved in more than one hierarchy, and sometimes more than one hierarchy is involved in the same function.

To conclude this chapter, something must be said about the relativity of the terms that will be used to refer to systems. Because the universe is a hierarchy of systems with many levels, the word "system" can be applied at any of these levels. The parts of any system will be referred to as subsystems, and the system to which the system under consideration belongs will be referred to as the supersystem. However, it must be remembered that it is only the context under consideration that determines the reference of these terms. A supersystem from one perspective is a subsystem from another, and a subsystem from that perspective a supersystem for yet another. For example, if one is considering atoms, particles are subsystems and molecules considered supersystems. If one is considering cells, molecules or organelles may be considered as subsystems and tissues as supersystems. Of course, subsystems have subsystems that have subsystems—and so on—and supersystems have supersystems that have supersystems. There should be no confusion as long as one clearly designates the level of systems under consideration.

These are some of the general characteristics of all systems, and in the chapters that follow we shall move from one level of systems to another in order to examine the applicability of these characteristics in relation to the present state of knowledge concerning those systems. In keeping with the idea of novelty, it will be necessary to add characteristics found in complex levels of systems but not in more simple ones.

Chapter 2

Simple Stable Order Systems

Throughout this book, we shall examine various levels of systems, from simple to complex, so as to bring them all into a unifying perspective. In this connection, a step-by-step schematic representation will be developed to serve as a skeleton around which to orient the many different areas considered. The first part of this model will concern the most simple level of system: energy, particles, atoms, and molecules. We will begin by introducing two basic terms: entropy and negentropy. These terms, especially entropy, have a wide variety of uses, some highly technical, but in this context we can be reasonably clear about those that are immediately relevant.

Entropy and negentropy are best understood in terms of another, probably the most basic, polarity, integration and disintegration. In every process of the universe, energy is used and released in the building of more complex systems, the transformation of systems, and the breakdown of systems. Basically, *entropy* refers to the degree to which the energy of a system is no longer usable. In order for something to cause change or motion, it must differ significantly in energy from its surroundings. Energy that "fades into the universe" does not disappear, but its level is so much the same as that of its environment that it has no "advantage" with which to accomplish anything. However, not all energy is released in an interaction. Much of it remains bound in simple stable systems from which it cannot be released under "normal" circumstances. Available energy is that which radiates at moderately high energy levels or is constrained in some system in such a way that it can be released and used in the breakdown of those constraints.

The absence of relations or constraints is randomness. Consequently, entropy, in some cases, has come to mean randomness. Until recently, there was no word for the opposite process, that is, for the building up of systems containing available energy. This process was first referred to as negative

Simple Stable Order Systems

entropy, which was then contracted to *negentropy*. The term was coined in connection with living systems, which must constantly capture and bind energy in usable packages so that they always contain "net" or excess energy. The second law of thermodynamics is that any system left to itself tends toward a state of greatest disorder, and the subsystems making up living systems do just that when left to themselves, that is, they disintegrate. A living system, as a whole, is so ordered as to have processes that equilibrate its disintegrative processes and maintain net available energy. In fact, the entire evolutionary process has been negentropic in building up ever more complex, unstable, and improbable systems. In the model we will be developing, therefore, there will be arrows to indicate these two opposing directions. Arrows to the right will indicate the (negentropic) buildup of complex systems, while arrows to the left will indicate the (entropic) disintegration of systems. The central line must be seen as a two-way street with processes moving constantly in both directions. The model will be referred to as *the developmental continuum*, and the first segment is shown in Figure 2.

Our consideration of *stable order systems* must begin with electromagnetic energy, although there is little to say about it in terms of the analysis of systems. As indicated in Chapter 1, all varieties of electromagnetic energy travel at the same speed (186,000 miles per second) but differ in energy, that is, in wavelength and frequency of oscillation. All varieties of electromagnetic energy have both wave and particle (photon) characteristics. However, there is no way to get "inside" a photon or an energy wave. Consequently, while

Figure 2
The developmental continuum.

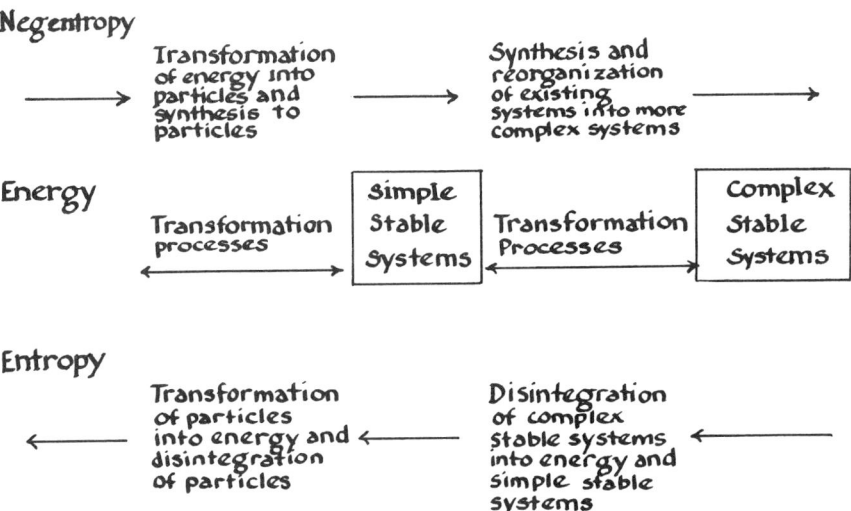

one can safely say that all systems are built up from this fundamental energy and the matter it forms, one cannot say whether energy is some kind of system in itself.

A similar problem confronts us with reference to the smallest subatomic particles. We shall assume that they are systems of energy in some form, but until more is known this is merely an assumption that fits the overall pattern developed here. As for the larger subatomic particles, they may be composed of smaller particles and energy in some systematic form, but the nature of those systems is still far from clear. In fact, "composed of" may be the wrong terminology, because mesons are not supposed to contain electrons, although electrons are produced when the mesons break down.

Scientists distinguish three classifications of particles, those considered light, medium, and heavy. With the exception of protons, which are stable, the heavy particles break down to lighter particles. One could assume that the process is reversible and that the heavy particles are some sort of synthesis of the lighter particles and energy. Whatever the inner nature of these particles, however, the situation is clearer for the next level of existence.

Atomic Systems

If we consider electromagnetic energy to be first level of existence, and subatomic particles the second, our first opportunity to apply general systems concepts in detail occurs at the third, or atomic level. The amount of information now available concerning the structure and processes of atomic systemic organization makes it possible to be much more specific. An atom is a system composed of energy, relations, and subsystems. Energy is present in the structure of the subsystems and in the relations among the subsystems. The subsystems are a nucleus and one or more electrons. In terms of the model we are developing, we must assume that an electron is a system, although at present, there appears to be no evidence for this assumption, which may prove wrong. On the other hand, there is sufficient evidence that the nucleus is a system whose subsystems are protons, neutrons, and pions. The systemic nature of the nucleus also becomes manifest in its equilibrating processes through which electromagnetic repulsion, which operates among positively charged protons, is balanced by the nuclear force. This nuclear force is currently described as an exchange process by which protons are changed into neutrons and back again by the transmission of pions from one to the other. This exchange process is so rapid that the normal repulsion among protons has no opportunity to disintegrate the structure of the nucleus.

This understanding of the nature of the atomic nucleus, together with the information that heavy particles can be broken down into light particles, suggests that it is reasonable to speak of nuclei and electrons as the subsystems of atomic systems. Atomic systems differ in a number of ways, but their systemic organizations always conform to the same structural constraints. The proportion of neutrons increases in the nucleus as atoms grow larger, but in neutral atoms the number of protons remains the same as the number of electrons, which are in orbitals around the nucleus. The number increases by 1 for each new element on the periodic table. The list of elements begins with hydrogen, which has one proton in its nucleus and one electron in orbit. The second element is helium, with two protons (and two neutrons) in its nucleus and two electrons in orbit. Looking at the periodic table of elements (Figure 3), one can see that the elements are numbered from 1 to 103, and that these atomic numbers represent the number of electron-proton pairs characteristic of the atoms of the various elements. However, the increase in the number of subsystems with increasing atomic numbers is not simply a matter of the addition of parts. The (probably novel) characteristics of nuclei and electrons are such that they relate in specific ways within atomic systems. The whole structure is based on the equilibration of the electromagnetic attraction, between positively charged protons and negatively charged electrons, and the kinetic energy of the electrons. Little is known about the arrangement of protons in the nucleus, but the arrangement of electrons provides a beautiful example of the exclusiveness and selectivity of systemic organization.

Electrons must be added in accordance with a specific pattern in the process of filling up to seven energy levels. When these energy levels were thought of as shells and orbits, they were indicated by the letters K, L, M, N, O, P, and Q. With the advance of understanding and the rejection of the Bohr model, it has become more common and useful to label the energy levels with principal quantum numbers. K has the property $N=1$, L $N=2$, M $N=3$, and so on. The advantage of quantum numbers is that they indicate the number of electrons atoms may have at a given energy level: The expression $2N^2$, where N is the number of an energy level, indicates the number of electrons that can occupy that energy level. Hence, energy level K, for which $N=1$, can have two electrons $[2(1)^2=2]$, energy level L, for which $N=2$, eight electrons $[2(2)^2=8]$, and so on.

Each energy level has from one to four sublevels, or subshells, indicated by the lower-case letters s, p, d, and f. The number of sublevels increases from 1 to 4 through the first four energy levels but, as Figure 4 shows, levels off at level 5 and decreases through levels 6 and 7. s sublevels hold 2 electrons, p sublevels 6, d sublevels 10, and f sublevels 14. Consequently, the K ($N=1$) energy level, with only one sublevel, can hold only two electrons. The L ($N=2$) energy level, with two sublevels, can hold eight. The M ($N=3$) energy level, with three sublevels, can hold 18. The N ($N=4$) and the O ($N=5$) levels, with four sublevels each, can each hold 32 electrons. The P ($N=6$) energy level, with three sublevels, could hold 18 electrons but only reaches 9 within the 103 elements on the periodic table. (Element 104 would add a tenth electron to the sixth energy level.) The Q ($N=7$) energy level, with only one

1 H																	2 He
3 Li	4 Be											5 B	6 C	7 N	8 O	9 F	10 Ne
11 Na	12 Mg											13 Al	14 Si	15 P	16 S	17 Cl	18 Ar
19 K	20 Ca	21 Sc	22 Ti	23 V	24 Cr	25 Mn	26 Fe	27 Co	28 Ni	29 Cu	30 Zn	31 Ga	32 Ge	33 As	34 Se	35 Br	36 Kr
37 Rb	38 Sr	39 Y	40 Zr	41 Nb	42 Mo	43 Tc	44 Ru	45 Rh	46 Pd	47 Ag	48 Cd	49 In	50 Sn	51 Sb	52 Te	53 I	54 Xe
55 Cs	56 Ba	57 La	58 Ce	59 Pr	60 Nd	61 Pm	62 Sm	63 Eu	64 Gd	65 Tb	66 Dy	67 Ho	68 Er	69 Tm	70 Yb	71 Lu	
			72 Hf	73 Ta	74 W	75 Re	76 Os	77 Ir	78 Pt	79 Au	80 Hg	81 Tl	82 Pb	83 Bi	84 Po	85 At	86 Rn
87 Fr	88 Ra	89 Ac	90 Th	91 Pa	92 U	93 Np	94 Pu	95 Am	96 Cm	97 BK	98 Cf	99 Es	100 Fm	101 Md	102 No	103 Lw	
			104 ?														

Figure 3
The periodic table of elements.

Atomic Systems

Shell	Number of Subshells	Label of Subshells
K	1	s
L	2	s, p
M	3	s, p, d
N	4	s, p, d, f
O	4	s, p, d, f
P	3	s, p, d
Q	1	s

Figure 4
Atomic energy levels.

sublevel, can hold only two electrons. Theoretically, levels 6 and 7 could hold more electrons, but the instability of large nuclei limits the size of atomic systems.

One might expect the addition of electrons to proceed in a regular manner from level to level, and so it does, but with some interesting complications. There is overlapping from the third (M) and fourth (N) levels on out, and the larger sublevels (d and f) have to wait until some of the higher energy sublevels have been filled. That is, before the d sublevel of the third (M) energy level can be filled, the s sublevel of the fourth (N) energy level must be filled. In the same way, before the d sublevel of the fourth (N) energy level can be filled, the s sublevel of the fifth (O) energy level must be filled. Figure 5 illustrates the order of acquisition, and indicates that f sublevels must also wait for the filling of sublevels in other energy levels. This order has an interesting result; that is, the outermost electrons are always in the s and p sublevels of some energy level. The two electrons of this s sublevel and the six electrons of this p sublevel constitute the valence shell of an atom. In every case, the nature of the atomic system is such that it determines where and in what order it will tolerate its various subsystems.

The atomic system is a simple, stable order system because it has specific, stable, and consistent patterns that determine its nature. It is stable because any major change transforms it into a different system, as we shall see in the subsequent discussion of the nature of ions and isotopes.

The first and second sublevels (s and p), constituting the valence shell of the atom, are the most important energy levels in determining the atom's external relations, or chemical properties. Ions are atoms with unequal numbers of (positively charged) protons and (negatively charged) electrons. The "excess" or "deficiency" of electrons is most commonly apparent in the valence shell. For each deficient electron, the ion carries one positive charge,

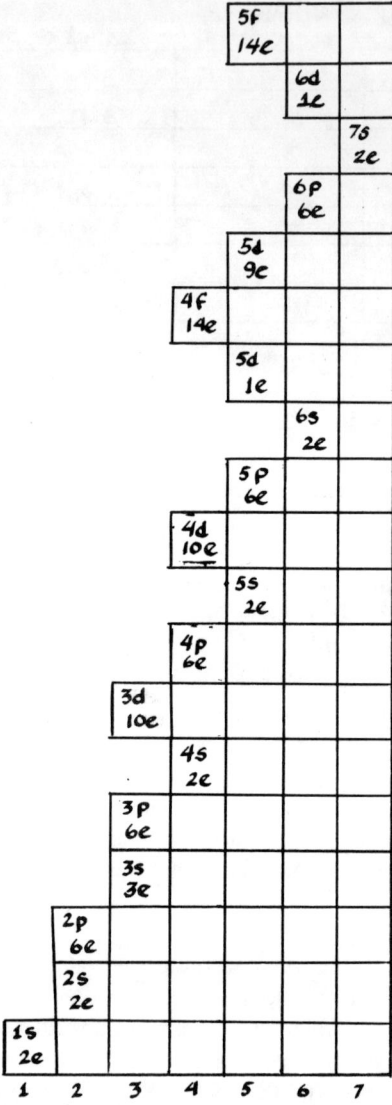

Figure 5
The order of electron acquisition.

and for each additional electron one negative charge. The properties of different ions and neutral atoms differ significantly. Are these systems the same or different? A change in internal relations in the form of the loss or addition of electrons does result in a significant change in external relations, but does that always make a different system?

Atomic Systems

In the processes of learning and adaptation, people often alter patterns of internal relations in ways that alter patterns of external relations, but we do not necessarily consider them to be different people or systems. Defining identity is a complex philosophical question. Its resolution may be a matter of convenience in classification and understanding. A person undergoing such unusual changes as brain damage or some other transforming experience may be said no longer to be the same person. What we mean is that that person has changed so significantly that we must find new ways of relating to him or her if we are to relate at all. This problem will trouble us at all levels of systems but, in the present context, we shall settle on a simple and somewhat arbitrary answer. Because electron exchanges are common and among the normal interactions of atoms, we shall regard them as adjustments of the same system. In comparison, extraordinary means are necessary for the removal or addition of protons. Such proton exchanges are normal only in places like stars, but in other circumstances they are very unusual.

The situation with isotopes is different and more easily decided. The isotopes of an element are atoms with different numbers of neutrons but the same number of protons in the nucleus. The mass of the isotopes differ but their external relations (properties) are affected very little. Different isotopes should probably be regarded as minor modifications of a single system because they occur frequently and perform chemically in the same way.

Beyond all this, however, a stable atom is a stable order system because, left to itself, its life expectancy is unlimited. (There are, however, radioactive atoms, which are unstable until they break down enough to become stable atoms.) The atomic system does not have to do anything to maintain itself. Transformations or adjustments in its system are brought about by encounters with other systems. In later chapters, we will be comparing *stable states* and *steady states*. The latter are different because, without the necessary interaction with other systems, they disintegrate into more simple systems. Their continued existence depends on processes within the system that constantly rebuild, energize, and maintain the system.

In Chapter 1, process was defined as the sequence of states of a system. All the systems discussed in Chapter 2 are in process. Referring to these systems as stable does not mean they are static or inactive. The system is dependent on the kinetic energy of the electrons remaining in balance with the energy of nuclear attraction. Within the nucleus, the energy of the mutually repelling protons is apparently balanced by the extremely rapid exchange of pions. If we could examine the particles that constitute the atom, we would probably find considerable activity within them as well. The atom may thus be viewed as a series of interrelated processes involving various kinds of energy and much motion and exchange. The system is preserved because the internal relations result in the equilibration of polar energies; that is, the particles and their energies are constrained within the system of relations. Taken by itself, the rapid motion of the electrons and protons would cause them to fly off into space. Taken by itself, the electromagnetic attraction of the nucleus would pull the electrons into the nucleus. The continuance of the entire system depends on the relation of autonomous parts of the system, that is, upon the maintenance of their identity and energy and the participation of those parts in the self-regulating relations that constitute the structure and processes of the system. For example, electrons must maintain their energetic motion, but within the specific energy levels determined by the nature of the system. The self-regulating nature of the whole system should be reasonably clear from some acquaintance with the order of electron addition in energy levels and sublevels briefly described above.

The system of the atom provides a good opportunity to elaborate on the idea that every system has boundaries. The atom has no skin or fence around it, but it has boundaries nonetheless. The stability of the atom requires that it have and maintain a specific number and arrangement of particles. In other words, the internal relations of the atom exist among specific member subsystems. A boundary is less a spatial division than a differentiation between what belongs to the system and what does not. This becomes more clear if one thinks of cultural boundaries. One may be in the geographical area dominated by a given culture and still not belong to it, for a cultural boundary concerns the adjustment of psyches. For the most part, the internal and external relations of the atom—and therefore its boundary—are clearly distinguishable.

Some apparently difficult questions arise in this connection. A neutron, because of its neutral charge, may pass "through" an atom without touching anything. While it is passing "through," is it within the boundary of the atom? Such questions are difficult because they are oversimplifications. The neutron would, of course, be spatially within the boundaries described by some of the orbitals but, in the absence of any relations, it would not be within the system of the atom. "Within" and "without" are matters of degree, and the distinction is often a matter of convention or convenience.

No boundaries are absolute, because there is always some transmission across the boundary. The most common of all transmissions is the equilibration of energy levels; that is, all systems tend to adjust to the energy levels of the surrounding systems. Beyond this, there is transmission of particles across the boundary in the formation of ions and isotopes and, in some sense, in the sharing of electrons in covalent bonding. The sharing of electrons involves us in the next level of system, molecules, and shall be considered in greater detail.

The atom, viewed as a synthesis of subsystems, has a structure and characteristics that are novel relative to its subsystems. As each new level of system is considered, new dimensions of structure are seen, such as electron clouds, which are capable of new patterns of external relations such as electrostatic and covalent bonding. This idea of relative novelty, combined with the idea of levels of systems, allows a picture of the universe wherein each level of system requires for its development the novel characteristics of all the preceding levels and incorporates them in its own relations. The novel structure of the atom, for example, makes possible the two kinds of bonding that make molecules possible. We shall see that it is the novel characteristics of molecules that make living systems possible, just as the novel characteristics of human psyches make cultures possible. In other words, the novel characteristics of systems make it possible for them to become subsystems in larger systems. In the next section, we look at atoms as subsystems in molecular systems.

Molecular Systems

There are two equilibrating tendencies underlying the bonding of atoms. The first is the tendency to have a full valence shell, the second the tendency to neutralize the electrical charges of this shell. By looking at the periodic table (Figure 3), we can see that relatively few atoms have full valence shells, only atoms in the extreme right-hand column: Helium with two electrons in its 1s energy level and the rest with eight electrons in the combined s and p sublevels of their outermost energy level.

Atoms that do not have full valence shells can come closer to this goal by acquiring or giving up some of their valence electrons. This results in ionization of the atom, that is, it will have a positive charge for each electron lost and a negative charge for each electron gained. This, of course, interferes with the equilibration of electrical charge unless the ion joins itself to another ion of opposite charge to form an electrostatic bond. Such bonding brings both atoms closer to full valence shells and equilibration of charges, and contributes to the formation of crystals.

Another manner in which atoms can satisfy the demands of these two types of equilibration is covalence, or the sharing of one or more electrons. Insofar as possible, electrons are paired in their orbitals around the nucleus, but unpaired atoms are often found in the valence shell. The covalent bond consists of the sharing of these unpaired electrons rather than, as in ionic bonds, the actual transfer of electrons from one atom to another. Nonetheless, covalence is just a different method of obtaining the same goals, that is, a full valence shell and electrical neutrality. A hydrogen (H) atom has one electron in its 1s valence shell (which holds only two) and is electrically neutral. Were it to acquire a second electron, it would have a full valence shell but would no longer be electrically neutral. When two H atoms enter into a covalent bond and form an H_2 molecule, however, each has a full valence shell and is electrically neutral. This is possible because the two electrons have paired up in a molecular orbital incorporating both nuclei in such a way that each nucleus is surrounded by the pair of electrons. At the same time, each nucleus has only a half interest in the total electrical charge of the pair of orbiting electrons, and the whole molecule remains electrically neutral. The fields of the two electrons may be diagrammed as in Figure 6. Usual symbolic representations of this kind of bond are dots to indicate the paired atoms or a short line to indicate the single bond formed by the paired electrons, for example, H:H and H—H. Molecules of water, ammonia, methane, and a six-carbon sugar are represented in Figure 7 by the latter representation.

These two kinds of bonding are the fundamental relations by which all molecules are constructed. The largest and most complicated molecules are formed by the addition of more and more atoms by means of such bonds. For example, sugars, starches, cellulose, fats, and the like are built up by the construction of carbon chains, including the bonding of sugar units (see Figure 7). A sugar unit is a monosaccharide, meaning one sugar. Monosaccharides can be bonded together to form disaccharides (two sugars) and polysaccharides (many sugars). Long chains of glucose make starches and, with some different bonding, cellulose. Three sugars with the addition of carboxyl groups and bonded to a glycerine molecule constitute a molecule of fat.

Amino acids are carbon chains or rings with a carboxyl group on one end and an amine group on the other. Glycine is a very simple amino acid, whereas tryptophan has a very complex middle structure (see Figure 8). Each of the over two dozen amino acids has its characteristic organization, but the ends are always the same. The carboxyl end of one amino acid can be bonded to the amine group of another in a peptide bond, and a long chain or double chain of them forms a protein.

The development from sugars to polysaccharides and from amino acids to proteins illustrates how molecules serve as subsystems or subassemblies for larger molecules. Just as carboxyl groups and amine groups can be chained together in various patterns, so the most complex molecules are constructed of subassemblies.

Several kinds of subsystems are used in the construction of nucleic acids.

Figure 6
A hydrogen covalent bond.

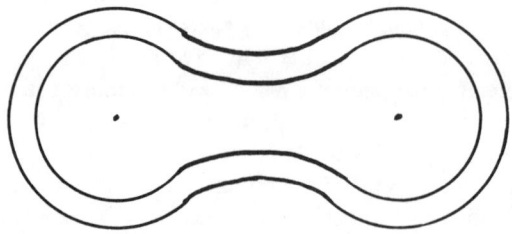

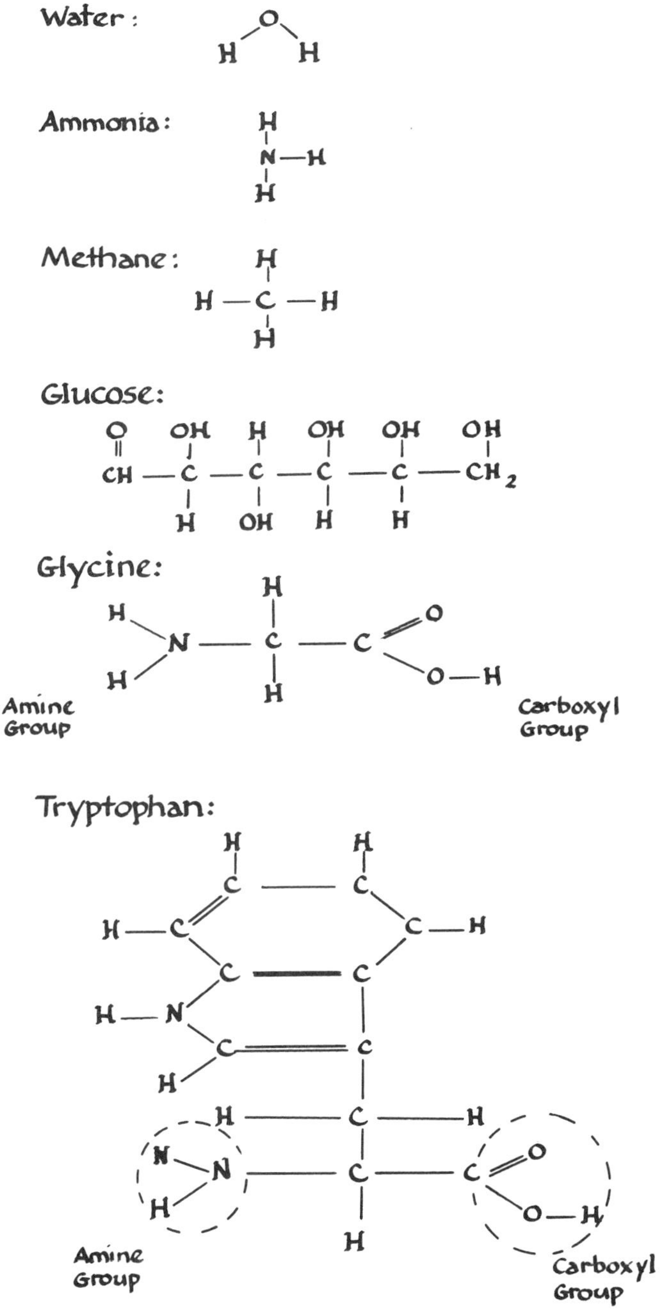

Figure 7
Representations of covalent bonds.

There are single carbon and nitrogen rings called pyrimidines, such as cytosine. There are also double carbon and nitrogen rings called purines such as adenine. When these systems are bonded to one of two five-carbon sugars, ribose or deoxyribose, they form nucleosides. Nucleosides are thus the next higher level of subsystem. A third necessary subsystem is a phosphate group. Joined to a nucleoside, it makes the next higher level of subsystems, nucleotides. There are eight major varieties of nucleotides used in making the nucleic acids, DNA and RNA (see Figure 8): Two kinds of sugar can be used and there are four varieties for each sugar.

Figure 8
Subsystems for nucleic acids.

Cytosine:

Adenine:

Phosphate group:

Ribose nucleotides:
  Adenine ribose phosphate
  Guanine ribose phosphate
  Cytosine ribose phosphate
  Uracil ribose phosphate

Deoxyribose nucleotides
  Adenine deoxyribose phosphate
  Guanine deoxyribose phosphate
  Cytosine deoxyribose phosphate
  Thymine deoxyribose phosphate

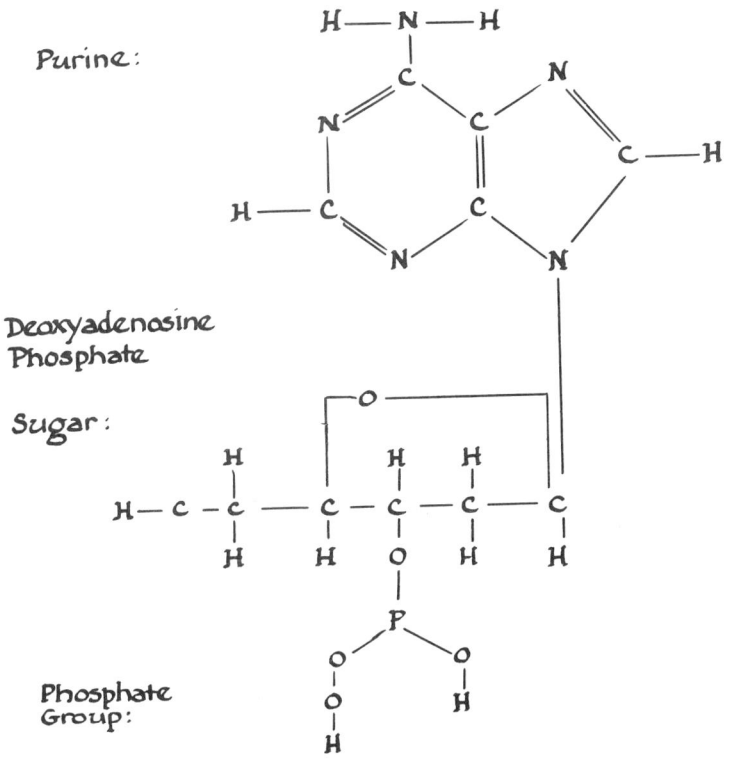

Sugar Phosphate	— Adenine —	— Thymine —	Sugar Phosphate
Sugar Phosphate	— Guanine —	— Cytosine —	Sugar Phosphate
Sugar Phosphate	— Thymine —	— Adenine —	Sugar Phosphate
Sugar Phospate	— Cytosine —	— Guanine —	Sugar Phosphate
Sugar Phosphate	— Adenine —	— Thymine —	Sugar Phosphate
Sugar Phosphate	— Guanine —	— Cytosine —	Sugar Phosphate

Figure 8 (*continued*)

Ribose nucleotides
  Adenine ribose phosphate
  Guanine ribose phosphate
  Cytosine ribose phosphate
  Uracil ribose phosphate
Deoxyribose nucleotides
  Adenine deoxyribose phosphate
  Guanine deoxyribose phosphate
  Cytosine deoxyribose phosphate
  Thymine deoxyribose phosphate

These are fairly complex molecules, but they are also the next higher level of subsystem on the way to nucleic acids. The nucleotides must be joined in pairs and then the pairs bonded in chains of hundreds or thousands and more. In a single fertilized human reproductive cell, there may be as many as 5 billion nucleotide pairs in the DNA molecules of chromosomes (see Figure 9). This highly complex system of nucleic acids that makes up the chromosomes is only one subsystem among many in a living organism.

There are other lines of development along which a variety of atoms are bonded together into molecules that can serve as subassemblies. For example, they can be strung together into long chains to form various kinds of fibers.

Figure 9
The developmental continuum.

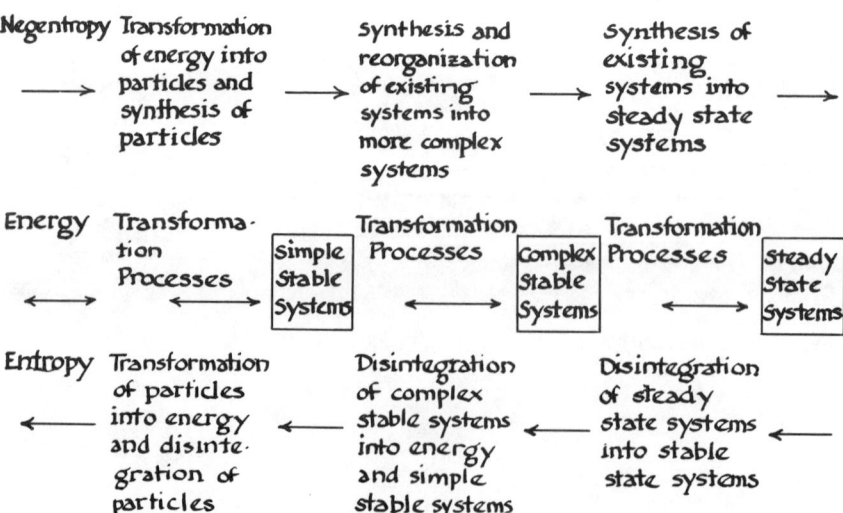

Molecular Systems

From such constructions come natural fibers such as silk or rubber and synthetic fibers such as nylon or dacron. In fact, the entire crust of the earth, including what grows, is made up of systems of atoms, crystals, and molecules built up into larger and larger systems. In this corner of the universe, it appears that the universe has been growing more and more complex systems in a long process of evolution. A slow but steady construction of new systems with novel characteristics has made possible the construction of newer and more complex systems.

All of the systems discussed so far, up through macromolecules, are stable systems. This does not mean that they will not disintegrate fairly rapidly. Some may start to decay as soon as they are separated from the living organism, but they must still be regarded as stable states. The difference between a stable and a steady state does not depend on the length of time a system endures. The distinction is between systems that have controlled input, output, storage, and maintenance processes, and systems that have repetitive processes and no reconstruction and that are, therefore, stable while they last. Some particles last for only fractions of a second but are stable systems while they last, while some living systems last hundreds of years and are steady state systems as long as they live. Of course, some simple stable state systems may last endlessly unless or until some environmental force causes them to change.

What has been said about atoms as systems also applies to molecules and macromolecles: They exist as systems of constrained energy through the equilibration of polar forces. The constraints are the relations among the parts that constitute the system at all levels of the system and its subsystems. In the case of crystals, molecules, and macromolecules, the relations are electrostatic and covalent bonds. These are the internal relations of the system, and its external relations are the ways it relates to those systems that are "outside" its boundary. The diagrams in Figure 8 indicate the specific structure of some molecules, and the abstract structure is indicated by the laws governing the possible patterns of bonding that determine the transformations from one system to another. Even at this level we are dealing with systems of systems of systems. Macromolecules are constructed of particles, atoms, and various levels of complex molecules. This will be the pattern in all that follows, so that the most complex systems will be seen as further stages in the gradual build-up of complexity and novelty.

Chapter 3

The Universe as a Physical System

At basic levels of organization the whole universe is made up of the kind of physical systems described in the preceding chapter. All of these systems and the universe as a whole appear to be constituted of energy and patterns of relatedness and levels of subsystems. This hypothesis is based on the assumption that the transition from electromagnetic radiation to particles that have rest mass is some kind of transition from kinetic energy to a pattern of internal relations. Consistent with the model we have developed and not in conflict with existing theories, a useful image may be presented of the smallest massive particles: Perhaps they are energy turned back upon itself in some way to form a vortex. Were this the case, then for these systems mass would be the inertia, that is, the resistance to acceleration, of the vortex.

However we conceive of the smallest particles, from larger particles on up, to be is to be related. That is, the only way to describe the existence of anything from the larger particles on up is to describe its patterns of internal relations, its external relations, the transmissions across its boundary, and the laws governing its transformations. How this idea incorporates a description of pure electromagnetic energy and the smallest particles is not known, but it is clear that all relations among systems involve some form of energy. The basic relations are attraction, repulsion, relative neutrality, and relative motion.

Relations among particles, atoms, and molecules involve all these basic relations. There are positively charged particles, negatively charged particles, and about as many neutral particles as charged particles, all in relative motion. Like charges repel, unlike charges attract, and the neutral particle is indifferent to the charges; neutrality is a state in which there is neither attraction nor repulsion. We speak of relative neutrality because, in general, systems are indifferent to some relations only: A system totally indifferent to all relations would be unrelated to anything and so nonexistent. The possibility of being constrained by electromagnetic relations depends on the proximity of the systems involved. Nuclei, atoms, and molecules appear to be constituted by equilibration processes among these relations.

Gravitational energy characterizes the universal attraction among systems with rest mass. This relatively weak force becomes significant as a relation only when the mass is large, or when systems are in close proximity and other forces are either absent or weak. It is proportional to the masses of the systems involved and inversely proportional to the square of the distances separating them. It is especially significant in the processes of the formation of stars and planets and in the interrelations of stars and planets in astronomic systems.

Nuclear energy is a property of the attractive power that binds nuclei together. It is very powerful over a very short range and is the source of the large amounts of energy produced in the stars. The force that holds electrons in their various orbitals *around* the nucleus is the electrostatic attraction between the positively charged protons of the nucleus and the negatively charged electrons.

Inertial energy is the energy of the relative motion of systems possessing mass. The motion of a system must be understood as relative, because it is not the same for all its relations. The inertial energy involved in any relation is a matter of the relative motion between systems. For example, the energy involved in the relation of two cars traveling side by side and moving slowly together is very small compared to the energy involved between these two and a car coming toward them. The law of inertia makes it clear that the motion of a mass is independent of some relations: A system having mass remains in motion or at rest unless relations with other systems having sufficient energy to overcome its inertia cause some degree of acceleration. Inertia may be defined as the characteristic of a massive system, a body, that causes it to resist a change in its motion. In this sense, the terms mass and inertia are synonymous and indicate that systems have an independence of motion relative to the amount of matter that constitutes the system.

The processes of the universe result from the interaction of these forms of energy, and therefore of the basic kinds of relations characteristic of these forms of energy. For general systems theory, the statement in terms of the relations has the advantage of greater generality. That is, the processes of the universe can be understood as the result of stress and equilibration among attraction, repulsion, relative neutrality, and relative motion. These relations exist in many different degrees and in an endless number of possible combinations, so equilibration appears to be a continuous process. A system in equilibrium or with equilibrating processes may come into relation with a strong form of energy and thus come under stress, requiring readjustment in the direction of equilibration. An atom is the result of the equilibration between the charges and motion of the nucleus and the electron cloud that surrounds it, and if it is subjected to sufficiently high levels of electromagnetic energy, kinetic energy (temperature or collision), or pressure, or the more powerful attraction of another atom, some of the electrons may be removed. The atom is ionized; that is, it now carries a charge. If possible, it then equilibrates by joining with a suitable ion of opposite charge and rejoining its electron. If the atom then falls into water, it is subject to a new stress from the electrical polarity of the water molecules, and the ions will be forced to equilibrate in a different direction by separating and relating to the water molecules. Chemical reactions are the result of stresses caused by an imbalance in relationships and a realignment of those relationships toward better equilibration.

This line of thinking suggests that equilibration and stress constitute another very basic polarity in the nature of existence. We have already mentioned the relative autonomy–system dominance polarity and the integration–disintegration polarity. The equilibration–stress polarity may seem very similar to the latter, but there is a difference worth noting: The equilibration–stress polarity allows for consideration of the degrees of equilibration and stress and the degrees of adjustment and modification. In other words, the processes of the universe involve not merely integrating and disintegrating systems, but often the adjustment, modification, and transformation of existing systems of equilibration in response to stresses. When the stress is among two or more systems in the form of attraction, it is equilibrated by means of a synthesis of the systems, as in the build-up of complex molecules. Perhaps more often, stress among two or more systems is equilibrated by the mutual transfer of subsystems so as to transform the original systems, as in chemical reactions. Sometimes, the equilibration is achieved through a distribution of subsystems, as in diffusion in fluids, sometimes through a distribution of energy, as in the equipartition of energy among molecules.

Most of the interactions under consideration in this chapter are processes involving two or more systems while, in Chapter I, "process" was defined as the sequence of states of a single system. However, the difficulty disappears if one considers the universe as a system. From this perspective, all processes are sequential states of some system, often of a series of levels of systems and subsystems. The universe is a self-contained system, and as such is different from other systems. It is meaningless to speak of a boundary to the universe as a system and, for the same reason, the idea of external relations has no significance. Without the concepts of boundary and external relations, transmission, input, and output also hold no meaning. However, all other concepts are applicable to the internal relations of the universe as a system, and therefore all processes can be considered sequential states of the universal system.

The irrelevance of the concepts of boundary, external relations, transmission, input, and output, and the nature of the universe as a system of internal relations, can be seen clearly in relation to the conservation laws. There are many kinds of transformations of systems and transformations between systems and forms of energy, but the content of the universe, whether matter or energy, remains the same: there is no other place for it to go than into relations and processes.

The law of conservation of mass–energy indicates that although mass and energy can be transformed into one another under certain conditions the total amount of mass and energy in the universe remains constant. Other conservation laws, governing various kinds of interactions such as the conservation of momentum, angular momentum, charge, parity, and spin, are also concerned with conservation of the mass–energy content of the universe through various kinds of transformations.

Systems, including the universe and all its subsystems, are equilibrating structures of relations and processes. More explicitly, they are constructions of relations and processes, because although most systems must be described as constructions of subsystems, relations, and processes, subsystems are in turn constructions of relations and processes. It is more accurate to indicate the presence of the subsystems, because their characteristics of subsystems must be present before the next level of system can be constructed.

We have seen repeatedly that systems are equilibrating constructions of opposing or polar relations and processes. Three basic polarities—subsystem autonomy–system dominance, integration–disintegration, and equilibration–stress—have already been described. Still other polarities may be involved in the equilibration of polar relations, including attraction–repulsion, attraction–resistance, and repulsion–resistance.

Equilibrating relations within a system tend to conserve the system; that is, all systems resist change to some degree. The degree, of course, depends on the strength of the binding relations and the system's adaptability to various stresses. We have defined inertia as the resistance of a system (mass) to change in motion, and the system may resist other changes, such as transformation as well. The processes of the universe can be described as a pattern of equilibration to stressful relations that overcome the resistance to change of the systems, and the subsequent equilibration as a result of the stress. Equilibration, in this broad sense, may mean the disintegration of systems at some levels and the release of mass and energy to more simple and stable forms. It may also mean transformation into systems of comparable complexity, the synthesis of more complex systems, or simply an adjustment or adaptation of the system.

Continuum of Differentiation

The very existence of distinguishable systems within the universe presupposes differentiation. If the universe is not a homogeneous blur, there must be differences that distinguish part from part. No matter how similar, distinct systems can be distinguished by the events affecting them, events that differ by their location in space and time. Even this simple level of distinction, however, requires that the system be recognizable as entities, or figures, some how separated from their environmental ground. This brings us once again to the idea of boundary and with it to the necessity of integration. A system exists because it has caused some of the content of the universe to "stand together" in a structure of relations and processes. Further differentiation may then occur through growth. At the level of physical systems, growth may come about through further integration of additional parts, as, for example, in the growth of crystals. Stress may give rise to further differentiation by precipitating adjustment or adaptation of the system. If the stress is strong enough to force a change but the system is incapable of appropriate adjustment or adaptation, the stress becomes a crisis and forces transformation and/or disintegration. We now turn to more complex systems, where additional possibilities arise.

Continuum of Uniqueness

The schematic representation in Figure 2 shows a change in the degree of uniqueness as one moves up the levels of complexity. That is, the uniqueness of individual systems relative to the general characteristics of a class of systems increases with the complexity of systems on the developmental continuum. At the level of particles, atoms, and simple molecules, class characteristics are uniform for all individual systems. However, as complexity increases, so does the likelihood of individual variations in structure within a given class.

For example, proteins are molecules composed of chains of amino acids. The over two dozen amino acids can be joined in any sequence, with no limit on the repetition of similar units. Consequently, the possibility of different kinds of proteins is almost endless. DNA and RNA molecules are another example. As described in Chapter 2, these macromolecules are composed of series of nucleotide pairs. These can be arranged in any sequence without restriction on the repetition of similar pairs. As in the case of proteins, this makes for almost endless possibilities of variations. The arrangement of these nucleotide pairs is related to the gene structure of chromosomes, which are the basis for individual differences in organisms.

In succeeding chapters, we shall see that this increase in individual variation continues with increasing complexity of systems until the case of human individuals and societies, where class generalizations are usually considered as falsifying the individual.

Closed and Open Systems

There is some variation in the use of terms among those who are concerned with general systems theory, and this is particularly true of the terms "closed" and "open." The distinction came about because L. von Bertalanffy, the founder of general systems theory, realized that the advance of the biological sciences needed a different model from that which had evolved in the physical sciences. Great advances were made in the discovery of the laws of physical nature by treating portions of it as closed (isolated) systems. Laws were discovered, for example, for bodies falling in a vacuum or accelerating on a frictionless inclined plane. Once these laws were discovered, other variables could then be introduced to suit the needs of the scientist. Living systems, however, cannot be abstracted from their environments without disrupting much of what is essential to understanding them as living systems. To survive, they must be in constant interaction with their environment, and many of the subsystems and processes of an organism exist to provide this vital interaction. For this reason von Bertalanffy suggested that such systems must be understood as open systems with directed and purposeful relations to the environment. As we shall see in Chapter 4, this is an important distinction, although the terminology does raise some problems.

Strictly speaking, all natural systems are in constant interaction with their environment and, thus, to some degree, are open to their environment. The important distinction is between those systems in which the interactions are necessary for processes of constant maintenance and purposeful behavior, and those in which they are not. There are different degrees of openness, and precise terminology is needed to indicate the important distinction between living and nonliving systems. However, using the terms open and closed may lead to the erroneous idea that some systems are actually isolated in nature. Some kinds of closed systems do exist. The universe can be considered a closed system, but most closed systems are conceptual. That is, they are either formal systems such as one constructs in logic or mathematics or interpretive models constructed to represent limited aspects of natural systems. The latter models include only those relationships necessary to a particular investigation; they have been closed to all other relations to keep them simple and manageable. This kind of conceptual abstraction has proven highly successful in many branches of science. However, in working with the more complex systems, one runs an even greater danger of distorting the nature of the system by such abstraction. The study of human individuals is limited and misleading unless it encompasses their active interaction with their environment. They are not just individuals. They are also subsystems in a social system that in turn is a subsystem within an ecological system.

What follows will not depend very heavily on these terms. Instead, the terms "stable state" and "steady state" will be used to indicate the difference between nonliving and living systems. Again, the survival of a stable state system is not dependent on interaction with the environment. A steady state system is one that must be maintained, that is, continually rebuilt, by means of what is acquired through interactions with the environment.

Description of a Specific System

In light of what has been covered, we shall now suggest a list of descriptions for delineating specific systems. The list is derived from the general characteristics of systems. We shall have to expand it in subsequent chapters to accommodate the novel characteristics of more complex systems. The list is not exhaustive, and there is no particular significance to the order in which it is presented. Those interested in a more rigorous and exacting approach to the description of systems are referred to George J. Klir, *An Approach to General Systems Theory*, and Mario Bunge, *A World of Systems*.

1. A description of the external relations of the system and the variations of these relations.
2. A description of the internal relations of the system and the subsystems for which these relations are external relations.
3. A description of the relations among the subsystems.
4. A description of the energies within the system and the processes by which the system acquires, distributes, stores, uses, and transmits them.
5. A description of the polar tensions among some of these energies which constitute the system and subsystems.
6. A description of the stresses to which the system is subject and its equilibration processes.
7. A description of the balance of subsystem autonomy and system dominance for the subsystems and for the system relative to the environment.
8. A description of boundaries, transmissions, inputs, and outputs for the system and subsystems.
9. A description of the states of the system and the transformation processes between the states.
10. A description of the exclusiveness, selectivity, and discrimination of the system relative to the constrained variety created by the system's pattern of relatedness.
11. A description of the system levels within the system and the relative novelty that results from the construction of subsystems and the system.
12. A description of the hierarchies that constitute the system.
13. A description of the system's resistance to change in terms of its patterns of relations, relative autonomy, and equilibration process.
14. A description of the system's level of differentiation and uniqueness relative to its class of systems.

These descriptions will not be equally important for all purposes. Our intention is to present as inclusive a picture as possible. Some items overlap, and the list can be reordered in a variety of ways. In their present form, the form most suitable for the systems under discussion, they are a checklist of the characteristics to be considered in the description of systems.

Chapter 4

Steady State (Morphostatic) Systems

Our next step on the developmental continuum takes us to steady state systems. Figure 9 enlarges the model presented in Figure 2 to include steady state systems. There is no sharp line between the two kinds of systems, just as there is no sharp line between living and nonliving systems. One of the best known examples of a system that bridges the differences between life and nonlife is a virus. In its dormant state, it resembles the stable state of complex molecules, but when it invades a cell, it begins many of the processes of living systems and reproduces itself from the materials available in the cell. However, aside from such transitional systems, one can make a clear distinction between those systems that are definitely stable states (e.g., most molecules) and those that are definitely steady states (e.g., cells).

Equilibration

A steady state system must maintain itself by acquiring energy and materials and by continually rebuilding or replacing the parts of the system. That is, the system is conserved through the continual change of its parts. In such a system, there must be continuous transmission across its boundary in the form of input of needed energy, materials, and information and sufficient output of various kinds to keep the system in relative balance. Output includes waste products from the processes of the system, products for the use of a supersystem, and information necessary to interaction among systems. Some output in the form of loss of energy is always absorbed in the general energy level of the environment. This chapter is concerned with what are generally understood as biological systems; later chapters will be concerned with psychic and social steady state systems.

Steady state systems are largely process systems. If structure in general is understood as a system of relations and ordered transformations, then the structure of living systems must be understood as systems of relations and transformations which constitute continuous processes. Most transformations affecting stable state systems change them from one system to another, but the majority of transformations in which living systems are involved are internal and related to the processes which maintain the system. There are many influences from the environment that, directly or indirectly, cause transformations within living systems. These transformations are most often adjustments within the system and are part of its normal processes. At times, however, such environmental influences cause transformation of the whole system, disintegration for example.

The processes and transformations constituting living steady state systems are the orderly transformations of the stable state systems described in preceding chapters. Living systems are the integration of stable state systems in processes of transformations that provide the living steady state with the materials and energy necessary for its maintenance. In other words, living systems are constituted of energy, particles, atoms, molecules, and macromolecules and their normal relations and transformation processes.

However, some characteristics of steady state systems are novel relative to the characteristics of stable states, even those which constitute the steady state system. These novel characteristics are the result of the integration, control, and equilibration of stable state relations and transformation processes within the system. This is not to say that all parts of living steady state systems are stable states, but all are ultimately constituted of stable state systems and their relations and processes. The central business of a living steady state is threefold: to grow, to maintain itself, and to reproduce itself. All of its parts (subsystems) carry on processes in which stable state systems are acquired, modified, used, and returned to the environment in order to carry out these three functions. The novel characteristics of steady state systems are precisely those characteristics pertaining to the controlled way in which systemic processes of growth, maintenance, and reproduction are carried on.

These processes require a continual supply of energy and, in the final analysis, all living systems depend on energy that comes directly or indirectly from the sun. How this comes about is very complicated, and the most the present context permits is a very brief summary. However, for the more specific development of general systems theory, the details are of great importance. The quantification of the details, wherever possible, is the goal of general systems theory. The process whereby solar radiation becomes available as energy for living systems is photosynthesis, a process carried on by some bacteria, some algae, and plants. All other living systems depend on these photosynthesizing systems for "packaged solar energy." The process begins when sunlight strikes the highly specialized chlorophyll molecules, which then release an energized electron into a series of chemical transformations. These transformations result in the production of two complex molecules: adenosine triphosphate (ATP) and phosphoglyceraldehyde (PGAL). In Chapter 2, a number of the nucleotides composing nucleic acids were described. One such nucleotide is adenosine phosphate, which in addition to its role as a component of nucleic acids also functions as the major energy source for all work carried on in living systems. For this purpose, it is present as adenosine diphosphate (ADP). During the process of photosynthesis a third phosphate group is added so that it becomes adenosine triphosphate (ATP), and this new bond contains a high level of energy. At the sites in the living system where work is to be carried on, ATP is broken down to ADP and the energy of the bond is thus made available.

However, ATP is not the most abundant product of photosynthesis. The other, more abundant product is phosphoglyceraldehyde (PGAL), a carbohydrate that can be used "on location" as fuel for production of more ATP or further packaged for export or storage as more complex carbohydrates (sugars and starches), or as lipids (fats or fatty acids). These more complex molecules are more suitable for transport and storage because they are less active. Because not all cells of a plant carry on photosynthesis, and because not all living systems have photosynthesizing cells, the transport and storage of these complex molecules is extremely important to the continuation of living systems. The only way nonphotosynthesizing systems can produce ATP molecules is by burning complex organic molecules acquired directly or indirectly from photosynthetic cells.

A photosynthetic living system, given water, minerals, and gases from the environment, can, through the use of "packaged energy," manufacture all the carbohydrates, fats, proteins, nucleic acids, vitamins, enzymes, and other complex molecules essential to the maintenance of its structures and processes. Nonphotosynthetic systems are much more limited, and must acquire "packaged energy" in a variety of forms from other living systems. For example, animals can produce many, but not all, of their needed complex molecules. Some vitamins, amino acids, fatty acids, organic nitrates, and organic carbon must be acquired in finished form. Given these building blocks, they can then modify them and manufacture the particular kinds of molecules necessary to their own structures and processes.

There are more sources of energy than those we have just indicated. Carbohydates are the principal, but not the only fuel used in the production of ATP. All parts of a living cell are burnable. In fact, they are being burned more or less all the time. The burning process takes no account of the degree of importance except for the novel characteristic of organization and control. With all of these varied and interdependent processes going on in a cell, there is continual equilibration and control for the whole system and at every level of subsystem.

Some living systems are single cells; others are much more complex. In the most complex systems, cells are integrated into tissues, tissues into organs, organs into organ systems, and organ systems into a complex organism. The need for equilibration at every level and in relation to the whole is absolutely essential for the continuation of the system. A living system never actually attains a state of equilibrium, but it must be constantly equilibrating, that is, correcting in that direction. What brings balance to one subsystem, however, may cause imbalance to another. There must be constant compromise and adjustment among a variety of needs at a variety of levels for the continuation of the system as a whole. However, this is not to suggest that the goal of a living steady state is to arrive at a state of equilibrium. As Buckley[1] has emphasized, the model of equilibrium states, from the physical sciences, simply does not fit steady state systems. Equilibration, in the sense used here, must be understood as the reactions necessary to avoid extremes rather than the attempt to attain an ideal balance.

Competition, Cooperation, and Control

This line of thinking brings us to another polarity, between *cooperation* and *competition*. Organization and control in a living steady state system requires cooperation among subsystems at every level, and throughout the whole system from the lowest to the highest levels. However, each subsystem and subsystem of a subsystem has its own autonomous problems of equilibration. In complex systems, such as animal endocrine, nervous, or digestive systems, there are a number of hierarchic arrangements that must cooperate but will, nonetheless, often compete for available energy and materials. This interrelation of cooperation and competition produces unavoidable internal stresses and necessitates hierarchies of controls to manage the constant changes and adjustments within the system.

Although the various control systems may vary from chemical processes to complex nervous systems, their basic pattern is fundamentally the same. There must be a part of the system which attains information concerning the situation to be controlled, another that determines the response to the situation, and still another that carries out the reaction. Such adjustments are not made by a single act or reaction. There must be continual feedback, that is, the information-gathering part of the system must take account of the new situation resulting from the reaction and report the new circumstance to the determining part of the system so that further adjustment can be made. This cycle must be repeated continuously until the needed adjustment is achieved. However, the parts of a living system do not arrive at some final or ideal adjustment.

The dynamic nature of this kind of control is easily illustrated by the thermostat that is part of the heating system in the average home. The

---

[1]Buckley, Walter, 1967. *Sociology and Modern Systems Theory*. Englewood Cliffs, New Jersey: Prentice-Hall.

thermostat does not keep the temperature of the room at the exact temperature set on its dial, say 70°F. It keeps room temperature within a certain differential, perhaps between 69°F and 71°F. When the room temperature falls to 69°F, the thermostat calls for heat, and when it rises to 71°F, the thermostat cuts off the supply of heat. In living systems, the needed adjustment must often be attained by successive approximations and continual feedback. For example, when you reach to pick up some object, the act is a series of small adjustments, each based on feedback reports. Ordinarily, we are so well organized that we are not aware of the process, but if you try to thread a very small needle, you may become aware of the problems of "overshoot" and "undershoot." That is, the action called for may be too much or too little. This results in oscillations, so that you have to stop, relax, and start all over again. Individuals often encounter this kind of problem when learning to drive or when driving a strange car. Their reactions tend to overshoot and undershoot until they "acquire a feel" for the brake, clutch, accelerator, steering, etc. In other words, in control systems, there are often "hunting" processes involving trial and error until approximately the right result is achieved.

Given this kind of constant dynamic control, necessary cooperation among many parts and hierarchies, and unavoidable competition, there can be no question of a living steady state system being in a state of equilibrium, although the processes of equilibration never cease.

Another reason a living steady state cannot be in a state of equilibrium and cannot be supposed to be trying to arrive at such a state is the nature of its energy system. Steady state systems must work at all times to overbalance their own entropic processes; that is, they must continually build negentropy. Because of the constant demand on and use of energy in control, maintenance, energy production, and other systems, there must be a constant backlog of net or stored energy. This is essential, because steady state systems are constantly subject to stresses requiring adequate supplies of energy in various places and at various times. This necessary flexibility of the energy supply requires adequate storage and distribution of "packaged energy" and, hence, the steady state system's balance is always "off center" toward an excess of energy.

Perhaps the most significant reason a living steady state system is never in a state of equilibrium is that it is never totally passive or at rest; it is always an active system. The nature of control systems requires that they are always more or less "fired up," that is, taking account of the situation even if there is no change or stress. This is true of internal equilibration processes as well as external equilibration processes directed toward the environment. The information-seeking aspect must report "no action," if nothing else, and so on around the feedback cycle. This readiness to act and the supply of net energy produce a restlessness in the living system that means that, lacking sufficient stimulation, the system will probe for stimulation in its environment or create its own, for example, curiosity and play. Such expenditure of energy is a part of the equilibration process; that is, the system may need to release energy if the situation has been too quiet too long. In any case, it is clear that, while a living system must equilibrate, its goal is never a state of equilibrium. The steady state requires, rather, a fluctuation between relative rest and relative activity, not passive waiting for a stimulus to which it can respond. Its own inner conditions are as much the cause of its activities as anything in the environment. Hence, as Koestler points out, a stimulus–response psychology can never adequately account for the complexity of human behavior.[2]

Equilibration is a constant and unending process in a steady state owing also to the many stresses to which it is subject at all levels of its organization. We have considered the need for energy, and of course the supply and distribution of the supply never rests at any fixed amount. There are continual adjustments between input and output of energy at every level and location in the system, as well as among input, use, and storage for each subsystem and for the system as a whole. The same is true for the supply of needed materials, and every fluctuation produces some degree of stress that must be relieved by equilibration processes. For living systems, the accumulation of waste products, changes of temperature, and degrees of acidity all produce stresses requiring equilibrating responses. The environment is the source of many stresses, such as threatening situations, actual injury, poisoning, and disease. A stress becomes a crisis when there is no available reaction for equilibration. In such situations, the system either acquires a new pattern of response or suffers some degree of injury or disintegration.

System Controls

The control system making all of this possible can be traced back to the chromosomes. As we have seen, the most important working part of the chromosomes are the nucleic acid, DNA, chains. In speaking of genes, one speaks of sections of these chains that function as patterns for certain synthesizing processes. Apparently not all sections function as genes and some overlap, so one cannot actually speak of chains of genes. However, it now appears that these DNA chains, directly or indirectly, determine the nature of all the parts that go into the construction of a living system. They replicate themselves in preparation for cell division, and they are the patterns for the synthesis of RNA molecules of various kinds. In turn, these RNA molecules become the patterns for the synthesis of various kinds of protein molecules. These various kinds of protein, combined with carbohydrates, go into the construction of the various parts of a cell, which in turn control production, use, input, and output for the whole cell.

However, to say that the DNA chains are the center of control is to

---

[2]Koestler, Arthur, 1967. *The Ghost in the Machine*. New York: Macmillan.

oversimplify the process and to misunderstand the nature of a system. In the final analysis, the whole system is the center of control. The determination of any part of the system is, to some extent, the condition of the whole system and the state of interrelation and interaction among all of its parts. There are, of course, what von Bertalanffy calls "leading parts" in "control positions" relative to some aspect of the system.[3] However, to forget that the controller is controlled by the system is to miss the true nature of the interdependence within the system. As mentioned earlier, science has made great advances by studying parts of the universe in isolation, but it is becoming more and more clear that steady state systems must be studied in process. It then becomes clear that control processes within a system comprise all the interactions going on in the whole system.

For example, there is a variety of slime mold (*Acrasiaceae*) that exists as independent, single-celled organisms. They reproduce rapidly by simple cell division until there are large numbers of them and then, at a certain time, come together in one place and organize themselves into a plant like structure. The individual cells specialize, according to position, to form a foundation or stalk, some becoming sporelike and blowing away in the wind to form new colonies. In other words, it is the system as a whole and their position within the system that determines the specialization of individual cells.

A clearer illustration may be the well-known results of cell transplant. If the transplant is made before specialization, the transplanted cell will specialize in keeping with its new location. Apparently, no special determination exists within the DNA molecules; they are the same for every cell in a given organism. The combination of position, relation to adjacent cells, and DNA determines how a given cell will specialize and function.

It has been suggested that one of the great failures in understanding human psychology has been the failure to understand the part that position in a social system plays in determining the psychic system and behavior of the individual. This is not to say that position in a system determines the individual, only that neither the individual nor the social system do all the controlling. The total pattern of interaction within the system, which includes all the controlling and determining patterns, is the center of control. Strictly speaking, we cannot even stop at the system, because every system is part of a supersystem or several supersystems.

The chromosomes determine the goals of development because they contain the basic patterns upon which all growth is based. However, these are only the broad limits within which development must take place. The manner in which the goals of development are achieved may vary due to specific circumstances. One of the fundamental concepts in von Bertalanffy's development of general systems theory is *equifinality*. That is, one of the novel

---

[3]von Bertalanffy, Ludwig, 1967. *General Systems Theory*. New York: George Braziller.

characteristics of living systems is that they can adjust the means for attaining developmental goals when necessary. Living systems are not limited to a rigid pattern of development or behavior. What they cannot attain in one way, they may well attain in another. Living systems would thus appear to be guided by purpose, or a developmental map. They are not simply genetically determined, because they adjust to circumstances. They are not merely adjusting mechanisms, because they have specific goals and limits within which they must develop. Koestler elaborates this point as the relation between fixed principles and flexible strategies.

The characteristic of equifinality, which here means arriving at equal ends by different paths or means, indicates a distinction between biological systems and other steady state systems such as psyches and societies. Buckley indicates this distinction by the use of two terms. The first is "morphostatic," meaning static or determined form. Biological systems are morphostatic and equifinal because they may vary the specific steps but strive for the genetically determined goals of development. The other term is "morphogenic," meaning form originating. Societies are steady state systems, but they do not have a predetermined form to fulfill. Different kinds of societies can satisfy the needs of individuals; that is, different social forms can carry on the same basic functions. Buckley consequently emphasizes that social systems create and vary their forms as they develop and are therefore multifinal. However, the fact that they have a variety of forms does not imply that they are without limits. They are determined within limits, as are all systems, but the determinations are not genetic. They are determined by the nature, limits, and needs of the individuals who compose them, by the environment in which they develop, by interactions within the system itself. The important point is that the purposefulness of biological systems is directed toward genetically determined form, even though there is some freedom in the means of achieving it.

The nature of the interlocking control systems of a living steady state system may best be understood in terms of the hierarchic structure of these systems. Their hierarchic nature becomes clear if we consider the systems that have been covered from Chapter 2 on. Living systems are hierarchies of organization of energy into subatomic particles, particles into atoms, atoms into molecules, molecules into macromolecules, macromolecules into organelles, and organelles into cells. For more complex systems, cells are organized into hierarchies of tissues, tissues into organs, organs into organ systems, and organ systems into organisms. The point of this review is that a hierarchy of controls must exist corresponding to the hierarchy of levels of component systems. The whole system actively determines to a significant extent how each part functions.

Let us now recall the basic polarity between subsystem autonomy and system dominance. Each subsystem and control subsystem possesses a degree of autonomy necessary to maintain the part of the system to which it is directly related. Yet each is sufficiently dominated by the interrelations of more inclusive subsystems and of the total system to carry out its proper functions for the benefit of the whole orgranism. Its own processes are modified to remain in balance with the processes of the whole system. These adjustments are neither perfect nor complete and must be seen in relation to various polar tensions within the system and subsystem.

Specialization

The existence of systems is directly related to the process of differentiation; that is, entities must be separate and individual before they can be integrated into a system. Differentiation in space and time is not enough; there must be differences in kind for there to be systems as they now exist. As we have seen, systems are sets of relations which equilibrate polar forces or processes; that is, systems are processes that equilibrate differences. Differentiation means that an entity or system has become something specific out of all that is generally possible. Among more simple systems, there are clear classes or kinds of systems, all of which have identical characteristics, that is, atoms of a given element. As more complex systems are synthesized from more simple systems, novel characteristics arise, and with them classification becomes less precise because individual differences become greater. In connection with living systems, this begins with protein synthesis. Carboyhydrates and lipids are fairly limited in terms of variety and are interchangeable among most groups of living systems. Carbohydrates are carbon chains, and differences are limited to the length of the chains and the distribution of hydrogen and oxygen atoms around the chains. Other differences are related to the kind of bonds, for example, the difference between starches and sugars. Fatty acids differ from carbohydrates in the presence of carboxyl groups at one end of the carbon chain, and fats differ in being bonded to a glycerine molecule.

Significant possibilities for individual differences within a class begin with systems composed of fairly complex subsystems. For example, whereas carbohydrates are composed of chains of identical carbon atoms, proteins are composed of chains of amino acids, each of which may contain over two dozen different units. To take apart a sugar molecule and rearrange the carbon atoms would make no difference because all the atoms are alike. A rearrangement of the amino acids in a protein, however, result in an entirely different protein. This is why proteins are unique or specific for the individual living system, and why such difficulties are encountered with organ transplants. The protein of the transplanted organ is "unrecognized" by the receiving system as "foreign" and rejected. The same holds true for nucleic acids, which are composed of complex nucleotide pairs. In fact, it is the variety of nucleic acids that accounts for the variety of proteins.

Thus, along the developmental continuum, novel characteristics emerge with each new class of systems, and the uniqueness or individual differences among the systems constituting the class gradually increases. It is precisely this potential for individual differences that makes possible the more complex systems we have been considering. In living systems, the various subsystems, for example, the organelles, are all made of the basic building blocks: nucleic acids, proteins, carbohydrates, and lipids. For the system to remain viable, the subsystems must carry out special functions, and this requires special organization and structure. The subsystems in a cell, from membranes to nuclei, have unique organizational structure and unique characteristics. As a class, cells have many common structures and many common characteristics, but there are also many differences from cell to cell: In a complex living system, cells specialize. This means that of all the characteristics and functions that are common to cells in general, the specialized cell has lost some and developed others far beyond the usual. Specialization is thus a further development of uniqueness and makes possible levels of complex systems that otherwise could not be attained. The cell that constitutes a single-celled organism is generalized, in that it must carry out all the functions necessary for life. The cells in complex organisms specialize and develop some of these functions to a much higher degree. Such cells may be organized into a tissue specialized for some function in the organism, a group of tissues may be organized into a specialized organ, and a group of organs may be organized into an organ system, which in turn has a specialized function in relation to the whole organism.

However, no subsystem is totally specialized. Thus specialization must be understood in terms of a continuum from generalized to specialized, and every subsystem must be understood as having some degree of specialization on the continuum. There is some correlation between this continuum and the polarity of subsystem autonomy and system dominance. That is, the more specialized a subsystem becomes, the more dependent it becomes on the system as a whole. A specialized subsystem is, in varying degrees, incapable of surviving outside the system. On the other hand, no subsystem is utterly without autonomy because it must be sufficiently independent to carry out its own proper functions. The controls from the system may call for a certain production, but the details of how that production is carried out are determined within the subsystem to a large extent. Specialization may be understood as differentiation in relation to the functional needs of a system. The more complex the system, the greater the degree of specialization required to sustain it. Thus, along the developmental continuum, specialization increases in the direction of negentropy in some kind of correlation with the increase in complexity and the increase of the dependence of subsystems.

All this points to another developmental trend in living steady state systems, progressive *system dominance*. We have been considering those characteristics that make living systems different from nonliving systems and, in particular, those characteristics having to do with *subsystem autonomy*, that is, with the degree of adaptability relative to the attainment of goals and purposes. System dominance must be understood as the reduction of subsystem autonomy in favor of more rigidly fixed patterns of behavior. That is, system dominance has to do with the increase in the degrees of determination that operate within a system in relation to some of its subsystems. The larger and more complex a system becomes, the greater the degree of system-ordered behavior required of the lower levels of the hierarchy. In other words, the larger and more complex the system, the less deviation in behavior can be tolerated in those functions that are vital to the survival of the system. This is an interesting principle with, as we shall see, varying degrees of applicability to psychic and social systems. For living systems, there is a polar tension between those forces working for conservation of the system in its existing form and those working for some deviant or innovative pattern. The idea of progressive system domination suggests that any deviation the system allows tends to appear at the higher levels of the system's hierarchy.

There is a further aspect of control systems that we have not yet considered, and that is *mapping*. Mapping is the ordering of one system so that its parts correlate specifically with the parts of another. Living systems require information concerning the processes and situations necessitating controls. For information received to have any meaning or significance, it must be compared with information already contained within the system. Information is information for a particular system only if it correlates to some extent with the information already mapped within the system. If there is no correlation, the information will be noise for that system. Through the process of evolution, the chromosomes of living systems have developed so that they contain mapped information concerning the development and functioning of the organism and some mapping of some aspects of the environment. When DNA chains replicate themselves in preparation for cell division, the nucleotides available in the nucleus are mapped to the DNA chain. When RNA molecules synthesize proteins, amino acids available in the cytoplasm are mapped to the RNA molecules. The DNA and RNA molecules thus contain the mapped information for the formation of the parts of the living system. Many living systems contain mappings of information regarding elaborate patterns of behavior in relation to the environment, such as the instinctual behavior of insects and birds. Other, more complex living systems must map many aspects of the environment and appropriate patterns of responsive behavior as part of their maturation process, for example, learning in primates and man. The adaptability that comes with the ability to acquire new or corrected mappings of the environment has been of great survival value to more complex living systems and is dependent on the specialization of cells into nervous systems and brains. This is not to say that more simple living systems do not learn, only that there is a correlation between the specialization of structures and the degrees of ability to learn.

Living systems would not have lasted long enough to make any significant difference in the universe were it not for another novel characteristic, reproduction. Individual living systems do not last very long as compared to most of the systems that constitute the earth, but living systems through reproduction have been as durable as any other system. Because living systems reproduce, and because those individuals that are best suited for survival tend to reproduce more than others, living systems have been capable of evolutionary adaptation. This process has accounted for the continued movement along the developmental continuum in the direction of negentropy and greater complexity.

Description of a Specific System

More could be said about the details of living systems, but, for the purposes of general orientation, this is enough to indicate the direction in which further elaboration would go. In Chapter 3 were listed some of the items that should be included in the description of any system, and those suggestions apply to biological as well as physical systems. We may now list some of the items that are particularly relevant to a description of living systems.

1. A description of control systems based on information, mapping, feedback, and directed responses.
2. A description of the input, modification, distribution, and use of energy.
3. A description of the input of adequate amounts and appropriate kinds of materials.
4. A description of the synthesis and rebuilding of parts of the system.
5. A description of equilibration processes among subsystems and within the whole system.
6. A description of the processes for equilibration of stresses from the environment.
7. A description of differentiation and specialization of processes.
8. A description of reproductive processes.
9. A description of adaptation.
10. A description of the active nature of the system (net energy and spontaneity).

There is no particular significance to the order of presentation in this list and there is some overlap among items: It is possible to revise this, like the previous list, in a number of ways.

Chapter 5

The Universe as a Biosphere

The universe as a system, excluding living systems, was considered in Chapter 3. Now we shall consider the universe as a system including living systems. It may be stretching the point to speak of the universe as a biosphere since our only knowledge of living systems comes from the earth. However, given the numbers of stars in galaxies and the numbers of galaxies in the universe, the probability is very large that there are other planets throughout the universe on which living systems have developed. If we accept the contemporary view of the manner in which life evolved on the earth, it appears that life inevitably develops wherever the conditions are right. The development of each stage can now be understood as the result of the normal properties or characteristics of the components involved. It is inevitable that, given the right conditions, certain chemical reactions will take place and certain kinds of bonds will form. As we have seen, newly formed systems or entities have new properties that make possible higher levels of synthesis when the right conditions are present.

All this suggests that the universe is a morphogenic (form-generating) system and that what has happened on the earth is one of the normal directions in which the processes of the universe move. That is, the development of more complex systems appears to be a polar tendency to the disintegrative processes of the universe. The second law of thermodynamics states that any system left to itself will tend toward the state of greatest disorder, or the state of least organization. On earth, this tendency has been more than balanced by the tendency to build up increasingly complex systems. This is because the earth receives a constant supply of energy from the sun and can therefore move in a negentropic direction. However, the sun is "running down," that is, using up its energy, in accord with the second law of thermodynamics, and the solar system will, with the death of the sun, follow the same path. Whether the universe is itself running down is a difficult problem, for it is building up by some processes while running down in others.

Current astronomic theory suggests that stars are running down while new stars are being formed through the condensation of interstellar gases. Whether the universe as a whole is moving toward the state of homogeneous distribution of its energy and matter (entropy) or whether this tendency is balanced by processes that are building up new energy systems (steady state) cannot yet be resolved. These possibilities are tied to astronomical theories about the origin and processes of the universe. The Big Bang theory describes the universe as having been, at one time, a huge egg of highly concentrated matter that exploded, resulting in our present expanding universe. A further elaboration of this view led to the Bang, Bang, Bang theory. This is a cyclical view according to which the universe expands to a certain point and then begins to contract until it forms the great egg. This, of course, is followed by another great explosion, and so on *ad infinitum*. Another view, which stirred up considerable interest a short time ago, asserts that the universe is a steady state, matter being created and destroyed at about the same rate, so that the content of the universe remains constant.

Given our present state of knowledge, is there any sense in referring to the universe as a biosphere? The universe as a whole, and most of its observable parts, are certainly not living. It is, however, the kind of system that, at certain times and in certain places, creates and supports living systems. To that extent, it is a necessary participant in any biospheres that do exist.

In some of its phases, then, the universe is a morphogenic system and among the systems it generates are biospheres. Given the energy of the sun, the universal processes here on earth have been negentropic in building up higher and higher levels of complex systems.

The Earth as an Open System

From this perspective, it can be seen that the earth and everything on it are, strictly speaking, open systems. The energy coming into the earth from the sun and the energy radiating out from the earth into space cause constant changes and disturbances in the balances of the total environment. The rotation of the earth, combined with the energy of the sun, causes constant change in the amount of energy coming into and radiating out from various parts of the earth. These changes, combined with the gravitational energy from the moon, produce constant motion and change in the atmosphere and in the water on the surface of the earth. These currents in air and water contribute to variations in temperature, cloud cover, precipitation, and drought. Every such variation is accompanied by equilibration processes and is the cause of stress on other parts of the system. Consequently, the earth, as a system, is constantly changing and equilibrating its changing conditions.

The more complex systems considered in Chapter 4 cannot exist for long without constant interaction with, and support from, the processes that constitute the system of the earth. The more simple entities of the physical universe may, under reasonably steady circumstances, remain highly stable for long periods of time. Changes in interstellar gas clouds and stars may be measured in billions of years, changes in the structure of the earth and the composition of its atmosphere in millions of years. However, a fatal degree of disintegration in living organisms subjected to severe oxygen deprivation, excessive bleeding, an increase in internal temperature of few too many degrees, or severe chemical, electrical, or psychological shock may be measured in minutes. Simple physical entities can, for the most part, maintain a stable state as long as the universal processes do not subject them to excessive stress through significant changes in conditions. Living entities, on the other hand, will if left alone disintegrate from stresses arising within their own systems. A living entity must be in constant and intimate interaction with the universal processes for it to continue to exist.

Here a fundamental problem arises. A living system cannot simply participate in the universal processes and survive. Its inner homeostatic needs are specific, and its interaction with its environmental processes must be selective. It must function with a divided loyalty or through the constant readjustment of two different systems, the external system, on which it depends for survival, and the internal system, which is the necessary condition of its existence. From cells on up, each living and social entity is in some degree a part and in some degree a whole. If it is a system, it is to some degree a whole. On the other hand, there are no systems that are not, in some sense, parts in larger systems. Therefore we must take care to distinguish degrees of dependent participation and degrees of independent wholeness, because all living and social entities represent some compromise between these polar extremes.

Degrees of Dependence and Interdependence

There is a general tendency for the more complex entities to require more energy for their maintenance. In addition to, and because of, this greater need of energy, the more complex the entity, the more dependent it is upon the larger number of processes in which it participates. This may be seen in broad perspective by considering three levels of existence: simple physical entities, plant life, and animal life. Simple physical entities need only be "left alone" to survive in a particular state, but plant life depends on the constant availability of mineral nutrients and gases from the earth and atmosphere and energy from the sun. Animal life is dependent on a steady supply of minerals from the earth and atmosphere, as well as a steady supply of complex organic molecules in which minerals and energy have been combined in special ways by other living systems. This progression in the degree of dependence suggests that the more complex entities must have more systems working for them if they are to survive.

The more complex the entity, the less direct and immediate its relations to the rest of the universe. Simple physical entities, for example, react directly and immediately to changes in conditions. Plants are in direct contact with the earth, atmosphere, and solar radiation but exercise some degree of selection and control over the interaction between themselves and the environment. Both animals and humans are highly mobile and directly tied only to the atmosphere. For the rest of their needs, they can exercise a fairly high degree of selection as to time, place, and desirable situation for their fulfillment. This latter circumstance has, no doubt, played a significant part in our image of ourselves as being so very independent. However, while animals in general, and humans in particular, possess great independence, both are nonetheless entirely dependent on a delicate balance of forces among the systems of the inorganic and organic worlds. Human beings are dependent on a complex system of systems, of systems, of systems, and so on. This means that the inner processes of living systems, including our own, cannot be adequately understood without seeing them as intimate parts of larger processes, which are in turn parts of larger systems, and so on up to the whole universe.

The apparent contradiction of these two observations arises from the use of general terms without sufficient limitation and clarification. It serves to illustrate the important point that, in terms of existence within hierarchies of systems, it is unlikely that an entity is "all" this or that, or that any description is wholly and simply true of any entity. If it is true that all systems are balances of conflicting forces that under changed circumstances move to alterations of the balance, that entities are generally constructed of various levels of simpler entities, and that the novel properties and abilities emerge at new levels of complexity, then we live in a universe where the actual variety included in the internal and external relations of any system is greater than any concrete statement made about it. Especially as they concern the more complex systems, such statements must be recognized as dealing with certain aspects of entities within certain contexts and at certain moments of time. When the multiplicity of relations is not taken into account and the actual limitations of a particular statement recognized, then one becomes involved in needless confusion, seeming contradictions, and apparent ambiguities. One can say that life is full of such enigmas and that more complex entities tend to become both more dependent and more independent at the same time, or one can say that, in terms of certain kinds of relations, complex entities tend to become more dependent while, in terms of other relations, they tend to become more independent and separate. We shall come back to this point in subsequent chapters because it is central to some of the major concerns of human beings and their societies.

Subsystem Autonomy and System Dominance

Despite all clarification, tension in humans between the tendencies of subsystem autonomy and system dominance poses a crucial moral problem. The fact that all living entities must maintain a balance between participation in environmental processes and individual needs becomes, on the human level, the problem of balance between social conformity and individual self-assertion. This point will be clarified in some detail as we proceed through the rest of the book, but now we must give attention to the circumstances from which the problem arises.

Humans are not only participating parts of a universal system of systems; they also partake of the basic nature of all parts of that system and must come to terms with the same existential conditions. To be sure, there are novel aspects when one reaches the level of human beings, but these are novel aspects of the same basic conditions. Greater complexity both necessitates enlarged dependence on a broader base of supporting systems and gives rise to novel properties, some of which involve greater independence for some aspects of the entity. This situation does not become a problem until it is encountered at the level of human beings. At this level it is central to such psychological problems as identity, individuality, meaning, and adjustment.

The solar system most likely condensed out of a gaseous cloud produced by the explosion of a huge star or stars. Because of its position relative to the sun, the earth condensed into its present general form with significant quantities of all the elements, especially those necessary to life as we know it. In the course of time, the elements of the earth, energized by the sun, began to produce more and more complex molecules and crystals, which eventually contributed to even more changes. Together these developments have led to the emergence of a complex hierarchy of systems. From this point of view, the earth appears as a vast reservoir of countless atoms, molecules, compounds, and mixtures of all the basic elements. Energized over eons by the sun and by forces at work in the earth and atmosphere, the smallest units have given rise to larger units and to new properties for even larger syntheses. Before living entities could appear and remain, the earth-sun system had to evolve massive systems involving the movement and distribution of various materials. Despite its vastness, the earth does not have an unlimited supply of material and, if this were not continually being replaced, many kinds would have run out long ago. Further, without the massive systems just mentioned, even the available material would not be adequately distributed for the maintenance of life.

From an evolutionary point of view, systems have developed to their present form through various stages. However, for the sake of simplicity, we shall consider them in their present form. The water cycle requires the most energy and moves the greatest mass. Through evaporation and the metabolic processes of living systems, great quantities of water are taken out of the hydrosphere and released into the atmosphere, where it condenses into clouds and falls to the earth as rain and snow. Because all living organisms depend on water as a basic nutrient, this constant spreading of water over the face of the earth is absolutely essential to the spread of living forms. Not only are living forms primarily made up of water, but the hydrogen derived from water is an important constituent of organic molecules and is involved in many metabolic processes. The water cycle has also played a significant part in the distribution of mineral nutrients through the dissolution of the earth's crust and the transport of the resultant particles. These minerals include sulfur, chlorine, potassium, calcium, manganese, magnesium, copper, iron, and the like.

One of the most interesting and involved cycles is really—and is sometimes treated as—three different cycles, each so involved in the others that we shall treat them as a single process. This is the cycling of carbon, oxygen, and carbon dioxide through the atmosphere and various forms of life. The structural description of atoms and molecules in Chapter 2 makes it clear that carbon is indispensible to life because it is the backbone of all of the complex molecules on which life depends. Sugars, starches, celluloses, amino acids, proteins, and DNA and RNA molecules are all modified carbon chains. Hence, without carbon, there would be no life as we know it, and possibly no life at all.

The atmospheric carbon that is the building block of every kind of organic molecule is available only in the form of carbon dioxide. Weisz suggests that 200 billion tons of carbon are used every year in the photosynthetic processes.[1] If not replaced, the carbon dioxide in the atmosphere would be used up in a few months. However, the cycle does go on because the carbon dioxide is constantly replaced from the respiration of living organisms through various kinds of combustion and, occasionally, from volcanic eruption.

Carbon dioxide is not usable by all living forms and, therefore, carbon in that form is not available to all living forms. Plants use carbon dioxide in the photosynthetic process and release oxygen into the atmosphere as a by-product. They use oxygen and release carbon dioxide in their own respiration but, left to themselves, would soon run out of carbon dioxide and suffocate in an atmosphere of oxygen. The situation is reversed for animals. The cycle is complete and stays in relative balance, because animals must depend on the "packages" of energy and materials they can obtain from plants. The carbon

---

[1]Weisz, Paul, 1963. *The Science of Biology.* New York: McGraw-Hill.

that an animal needs is already incorporated in the organic molecules mentioned in Chapter 4. The life processes in animals take these molecules, using some as energy sources and modifying others to their own specifications. This means that animals are dependent on internal chemical processes for all their energy and must therefore "burn" quantities of organic molecules as fuel. Carbon dioxide is one of the byproducts of these processes and is thereby returned to the atmosphere to be used again by the plants. The metabolic processes in both plants and animals involve free oxygen and, therefore, the balance tends to stay relatively stable.

Another cycle that illustrates the high degree of interdependence of all living things is the nitrogen cycle. Although nitrogen is available in the atmosphere, it cannot be used by the vast majority of living entities. In fact, the whole system of living things is dependent on a group of tiny bacteria and blue-green algae that excrete nitrate ions into the environment. Some bacteria draw the free nitrogen from the atmosphere and incorporate it into their own organic molecules. When they die and decay, they contribute to the amount of ammonia (NH_3) available in the environment. Other bacteria use the available ammonia as a nutrient and excrete nitrite ions (NO_2^-). Still another group uses nitrite ions as a nutrient and excretes nitrate ions (NO_3^-), the form in which nitrogen is available for plants. Plants take up the nitrate ions as nutrients and transform them into NH_2 groups, as in amino acids. In this form, they can be utilized by animals.

The entire system of life, then, is a process of borrowing and returning elements and compounds from the earth and atmosphere in order to build the structures that, for a brief time, constitute a living entity. What we call death is nothing more than one phase in the continuous cycle of cycles that brings about the endless integration, disintegration, and reintegration constituting the living system in process. Within the individual organism, and throughout its life, the same cycle must be maintained. Parts must be built up, destroyed, and replaced for the entity to survive. There is no static stable existence for living organisms. The existence of a living system can only be maintained by the continual expenditure of energy in building, tearing down, and rebuilding internal structures and in maintaining a balance of processes. The more one learns about the nature of living things and their place within the living system, the more it becomes clear that every living thing is "of the dust," that "to the dust it must return," and that its "ashes" become the constituent parts of another living system again and again. If the endurance of living matter had depended upon the survival of any particular organism, it would have been nothing more than a momentary oddity happening now and then but of little consequence. The living system does depend on the survival of particular entities but only until they can reproduce themselves. Then, it depends on them to return their parts to the common supply for the use of other organisms and future generations. Hence, the whole living system is a balance of the use and replenishing of available materials.

Levels of Systems

The whole living system is also a pyramid of levels of entities with the higher wholly dependent on the lower for their existence. Smaller, more simple organisms must not only reproduce their kind but also be available as food for higher, more complex organisms. The body of the simple organism contains the materials the more complex organism needs to maintain its own system to survive and reproduce its kind and, in its turn, become food for another larger, more complex organism, and so on. Organisms are factories for producing the complex building materials needed by other organisms. Consequently, the living system is more than a balance; it is also a hierarchy of dependence and support from lower to higher.

The same pyramid of levels exists within the organisms themselves. Even the most complex organism is a system of cells and depends on the universal functions of cells for its survival. It also depends on tissues of cells and the organization of tissues into organs, and organs into organ systems. Thus the inner systems of the most complex organisms contain the whole range of levels of living organisms as their constituent parts. Such organisms survive not only by eating various levels of organisms but also by maintaining comparable structures within their own inner systems. In other words, the higher depends on the survival of the lower. If lions kill too many deer, they starve until their numbers decrease and the deer become more plentiful. If the deer eat too many leaves, they starve until their numbers diminish and the supply of leaves grows back. The same is true internally. Tissues cannot survive unless their cells do, organs cannot survive unless their tissues do, and so on. The survival of a particular organism depends on the survival of its constituent parts, the maintenance of an inner balance of processes, the survival of all the organisms on which it depends for the preparation of nutrients, the continued balance of and successful interdependence within its ecosystem, the continued balance of processes and interactions within the biosphere, and so on for ordered processes throughout the universe. Said more simply, the survival of any particular organism depends on the controlled balance of internal processes by means of successful interaction with the systems in its environment.

Competition and Cooperation

Before we move on to human beings and their problems, we shall make some general observations bearing on the relation of dependence and independence. Every organism is both intimately one with its environment and separate from it: No matter how separate, every organism is a vital part of the environmental system and cannot exist apart from it. The environmental system is equally dependent upon the proper functioning of separate entities—not upon any particular entity, but upon the proper functioning of sufficient numbers of them. This means there must be some balance between competition and cooperation, between participation and individual fulfillment. Every organism must be able to defend itself against destructive agents within the system and to compete successfully for available nutrients and energy sources for it to be able to fulfill its cooperative role in preserving the living system as a whole.

We have considered individual uniqueness in our discussion of proteins, DNA molecules, and genes. No matter how unique an individual organism may be, it is either the product of and an integral part of a system or else is, or soon will be, nothing. This facet of our model of life has important implications for the model of human beings and their understanding of their situation. From this point of view, an entity without an interacting environment is, or is on its way to becoming, a nonentity. We shall see in later chapters that this perspective appears to apply equally to everything from amoebas to people to ideas. Whatever one may mean by "independence," "individuality," or "doing one's own thing," the terms inevitably refer to something produced by, deriving its meaning from, and depending for continued existence upon environmental systems. This is not to deny the uniqueness of individuals. On the contrary, this perspective requires the recognition of genuine uniqueness at many levels as well as the recognition of genuine novelty. What it does deny is that uniqueness or individuality is ever produced by or maintained by an isolated individual. This is equally true of ants and architecture, of foreign policy and fish, of sand fleas and symphonies.

Selectivity and Discrimination

The problems we encounter in daily existence have their roots in patterns common to all living things. A human being, like every living organism, can never be either just a part or a wholly unique and separate individual. We are always both to some degree, and the tensions, frustrations, and threats encountered in the struggle to balance these polar extremes are at the very heart of our moral problems.

There is no clear line separating living and nonliving entities, but the novel kind of existence we usually associate with life and its problems comes into clear relief at the level of cells. For all systems, from cells on up, living is a full-time occupation. Systems, such as viruses, that are somewhere between life and nonlife, may function, cease to function, and function once again. For most living entities, however, to cease to function is to cease to exist. Therefore, the problem for living organisms begins with the semipermeable membrane that is characteristic of all cells. Constructed of layers of protein with a layer of carbohydrates between, this membrane not only separates the cells as particular entities from their environment, but also marks the beginning of active selective participation in that environment. We now know that the cell membrane is not just a passive filter. It actively selects and passes, by means of chemical reactions, much of what is allowed in and out of the cell. The DNA molecules in the genes determine the nature of the RNA molecules, which in turn determine the nature of the proteins and enzymes. They in turn help to determine the nature of the structural elements that constitute the various parts of the cell. Similarly, the specific proteins and carbohydrates that constitute the cell membrane determine what kinds of entities will be passed through.

Although some of the environment must flow in and out of the cell for the cell to survive, the amounts and the kinds of entities must be selectively controlled. In this sense, then, the cell is an extension of the environment. At the same time, the cell must be an exclusive entity. From this basic level of life on up, the survival of an entity presupposes some degree of discrimination in relating to other entities and some degree of self-regulation. As is true of any enclosure, the cell has a limited amount of available space and, consequently, "too much" and "too little" are constant problems requiring continual adjustments. There must be adequate control of the content of water, the basic solvent, to provide needed hydrogen and oxygen, to aid in some chemical processes, to aid in the transport of needed materials around the cell interior, and to prevent the cell from dehydrating or bursting. In addition, there must be a ready supply of all needed materials, such as nitrogen bases, amino acids, carbohydrates, and enzymes of various kinds. Too much of one means too little of another and will interfere with the life functions of the cell. Accurate discriminatory selection and control are fundamental necessities for the survival of any living system. For most organisms, these functions are carried out by genetically determined patterns but, as one moves up the levels of increasing complexity, learning more accurate discriminations and more reliable generalizations comes to play an ever more important role.

The words we have just been using are intimately connected with some of the most trying problems of individuals and human society. Words like "exclusive" and "discrimination" imply negative as well as positive judgments that have undesirable connotations for many people. The present context is less complicated, although relevant to all living systems: Our model suggests that some adequate pattern of internal relations is necessary to the existence of any kind of entity. A pattern of internal relations maintained constantly in relative balance is vital to the existence of any living entity. This balance must be controlled by a constant inflow and outflow of elements from the environment. What is taken in should be useful in terms of—and certainly not destructive of—the existing patterns of the entity. There is always that in the environment that, if allowed in, will destroy the order by which the entity exists. Thus survival depends on the ability to discriminate the useful from the harmful, as does the preservation of any system or order in a particular state. The problem is further complicated by the fact that living organisms and their environments sometimes change and that what is useful or harmful varies according to the different stages.

For most of the living world, this is a matter of genetic determination or some variety of conditioned learning. For humans, it is somewhat different and creates special problems. We do not have much, if any, overt behavior that is genetically determined in any elaborated form. Our discriminatory judgments have been learned from the particular society in which they have evolved and in which they have been self-imposed. We have no reliable knowledge of our full potential in terms of development, although we have explored any number of ideas about what direction we should move in. Which kinds of ideas, associations, changes, and so on lead to healthy growth and development, and which are destructive? Before this can be answered, what kind of person and society are most desirable? Simply knowing that there are no easy answers to these questions leads to some significant observations. We have always evolved, in human societies, psychosocial patterns for survival and satisfaction, and quite naturally assumed these patterns to be the right ones. We have also always experienced an intuitive dread of changes that might lead to psychological or social disintegration. Thus we are inclined to discriminate negatively against anything strange or unusual. Some of the psychosocial environment must flow into us and out again for us to survive as individuals in a society. However, if what flows in and dynamically reacts is too inconsistent with the existing patterns, there will be change and possibly some destruction.

Education and religion have been and are still being used both to support existing form and to encourage change. How much change, what kind of change, and what will produce the right kind of change are all very difficult questions. The problem of determining what our psychosocial nature ought to be, and of maintaining ourselves and our society in a desirable form once attained, comprise a main source of our most difficult problems. There must be order and discriminatory judgment of some kind if individuals and society are not to disintegrate. Equally important, new form comes out of the creative synthesis of existing forms. From subatomic particles to international relations, new form never comes where there are no forms from which it can be synthesized.

Control Systems

All that we have considered thus far points to the fact that the survival of living entities necessitates systems of control, systems found in every living system, no matter how simple. The control system of the most simple living entity is extremely complex. Further, the systems of control follow the same hierarchy of development and interdependence we have already observed concerning the structure of organisms. To put this part of the model into proper perspective, we must go back and pick up one more thread, that of energy and energy sources. The basic source of energy for living systems on earth is the sun. Without the sun's energy, the earth would still be a reservoir of elements and compounds but would be relatively inert. Other energy sources, such as lightning, may well have played an important role in certain primeval stages of development, but from cell life on it is the sun that has made the living system possible. But direct sunlight can destroy living forms; it is modified sunlight that makes life possible. Even too much modified sunlight can be destructive. The particular layered structure of the relatively heavy atmosphere around the earth filters the light. The atmosphere acts as a blanket for the earth and saves the earth's surface from extremes of heat and cold. Without it, the sun would burn living entities out of existence by day and the cold would freeze them into extinction by night. Life is possible because the earth's atmosphere balances the amount of energy (temperature) and eliminates harmful kinds of energy (high frequency electromagnetic radiation).

It has been emphasized that life consists of processes that maintain living structures, these processes requiring energy. In addition, the basic elements and compounds that constitute living things are themselves complex and relatively stable systems of "packaged energy." Their continued existence is dependent on the strength of the internal bonds and the influence of external forces. However, as we noted in Chapter 4, development beyond macromolecules brings about a gradual shift to steady state systems, requiring a continual supply of new energy for the system to be maintained. Hence, a living system is not just part of the environment but also dependent for survival on its ability continually to incorporate energy from the environment into its own being.

All processes of living entities are, or are dependent upon, chemical reactions, and all chemical reactions either take up or give off energy. Hence, a balanced control of energy within the organism is absolutely essential. There must be a constant source of energy within the living system, and the release of that energy must be controlled within very narrow limits. Otherwise, the organism would burn up from the inside out. Therefore, the sun's energy must be made available in storable, controllable units within the organismic systems. The more complex the system, the more likely it is to possess more complex systems for controlling internal processes and patterns of behavior. Progression in the structural complexity of the organisms that have evolved brings with it a corresponding progression in the complexity and increased capabilities of the control systems possessed by these organisms. What we have said of the interrelated and hierarchical nature of various structures can be equally applied to control systems. These systems exist in hierarchies, in nature in general and within the individual entity. From matter to life, there is a gradual transition into novel kinds of systems, the novel differences increasing as the capabilities for control and selectivity increase.

Now we shall incorporate some of the characteristics of the physical universe. The general presence of equilibrating systems was discussed in earlier chapters. Equilibration refers to processes of systems in which opposing forces are balanced. Changing conditions, however, may produce stresses on a system and a resultant imbalance of forces. The term "system" is a relative term in that it depends upon the point of view of the person using it. In the final analysis, everything is part of one system; but, for practical purposes and for purposes of human understanding, parts must be isolated for separate consideration. When one speaks of a system being subjected to stresses as a result of changing circumstances, it must be remembered that those circumstances are processes working in systems that are acting on other systems, acting on still other systems, and so forth.

Whatever the nature and scope of the system under consideration, certain processes are involved in its movement toward a new state of equilibration. For example, consider the equalization of pressure between two adjacent spaces, the equalization of temperature (kinetic energy) among physical entities, the equalization of the distribution of molecules and particles through diffusion, the equalization of the concentration of water through osmosis, and the tendencies toward aggregation or disintegration under the right conditions. These, of course, will be recognized as the tendencies described more carefully in various physical laws. However, we need not get into a detailed consideration of them to show that physical systems tend to maintain relative states of equilibrium by such processes. Our primary concern is to notice the difference that develops in relation to living systems, a difference having to do with a degree of internal control and selectivity relative to the changing environmental circumstances. The transition from physical systems to living systems is a gradual movement from simple stable equilibrating systems to ever more complex controlled steady state systems. Many physical systems can be considered controlled systems in the sense that they are held in balance through most normal changes in the environment. The situation is different and more complicated for the control of constantly changing steady states.

This difference can be readily seen by considering that the semipermeable membrane of a living cell functions partly on the basis of diffusion and osmosis as would any similar physical system. The added difference is that the cell membrane actively selects and passes, by chemical reaction, some of the particles which enter and leave. If the membrane acted only as a simple filter, the living processes would be difficult or impossible because there would be little control over the content of the immediate environment. The roots of plants serve as a good illustration. If the root cells were simple physical systems, then, when the mineral content of the ground became more concentrated than that in the root cell, both cell and the plant would lose needed water. However, because of the metabolic activity in the cell, more water is taken in than is lost through osmosis, and the proper water pressure is thus maintained. Put another way, the difference is that the living system expends energy constantly in the attempt to selectively control its interaction with the environment. From the level of the cell on up, a careful balance of inflow and outflow in relation to many complex chemical reactions in which molecules and macromolecules are being built up and torn down continuously is maintained through the expenditure of energy in active control. This requires a system of controls that increases in complexity relative to the complexity of the organisms. No matter how elaborate the structure and control systems of an organism may be, it is dependent on the proper functioning of each of the lower levels of which the organism is constituted. There are controls that operate on the level of cells, tissues, organs, organ systems, and organisms as a whole. Thus, for example, the brain and nervous system of man are dependent on and involved in the successful functioning of all levels of control from cells on up.

With the advent of self-regulating systems of controls in living systems, equilibration is called homeostasis, that is, the tendency of a living system to maintain inner stability, in response to stress, through the ordered responses of its parts and processes. Survival ceases to be something that happens due to conditions. The living system can do certain things to modify its relations to environmental conditions, and thus enhance its chances of preserving its form of existence, because it can do specific kinds of work (use energy) to modify its interactions with the environment. Levels of capability follow levels of complexity, and the more complex living systems develop more complex systems of control with greater capability for selective interactions with the environment. The higher the complexity of an entity in the hierarchy, the greater the number of possible reactions to changing conditions. Here, the gradually emerging novel differences we have been considering appear in another light. Simple physical systems are predictable to a high degree because their reactions are simple, lawful, and direct. The progression toward more complex living systems makes prediction more and more difficult because of the larger variety of more complex reactions that become possible. This does not mean such reactions are less lawful and therefore unpredictable. It does mean they are less simple, less direct and, therefore, much more difficult to understand and predict. Consider, for example, the difference between predicting the behavior of a cake of ice, an amoeba, and a human being.

To some extent, therefore, living entities must take account of conditions and "decide" on the appropriate reaction. The use of quotes in the last sentence is intended to warn against taking such a term in any human sense. Our logic and language do not enable us to express degrees of characteristics or abilities. Were we able to express degrees in a form like mathematical powers, we could, for example, say that planarians had awareness cubed and humans have awareness to the ninety-second power. Such a system is not unfeasible but, for the present, the best we can do is show that ever-increasing evidence points to gradual increase in degrees.

The gradual increase in the ability to be aware and react out of an ever wider variety of "choices" brings us to the fundamental problem of human beings. Here, again, we could profit from a quantitative system for indicating degrees of separateness. Then we might say that simple physical systems have a separateness to the first power because they react directly and immediately with the environment, while humans have a separateness to the hundredth power. The increase of separateness is due to a time factor; that is, the greater the capability of the control systems of an entity, the greater the possibility of delaying reactions and even avoiding certain reactions altogether. Even an amoeba can learn to move away from danger to some extent, and higher animals do many complicated things to save themselves from the necessity of interacting directly with environmental change.

Each is reacting to the environment. The significant difference lies in the ability of the complex living control system to "take account" of environmental conditions and react to them before a reaction is forced upon the system. When conditions change, a simple physical system must suffer the consequences, but a bird will migrate to a place where conditions are more favorable. As we consider the level of human beings, we find that their ability for creative synthesis of new solutions has raised the possibilities for modifications of interactions to a high degree. This is most readily apparent in technological advances, but these have only been possible because of psychological advances. Logic, rational thought processes, and scientific method have made these advances possible, but they have emerged in humans slowly and in proportion to their ability to separate their higher brain functions from the more immediate and subjective interaction processes.

Our central problem has always been that we are caught between the polar extremes that characterize the universe and are thereby frustrated and threatened. Applying general systems theory to humans, one is bound to repeat old truths within a contemporary perspective. As finite limited creatures, we participate in and are limited by the nature of the systems of systems of which we are an integral part. From the electrons that make thought possible to the atoms and molecules which constitute our every part, we are subject to the same lawful processes and limitations that are characteristic of the whole universe. In two important ways, however, we can transcend these conditions to some extent. Within limits, our technological inventiveness enables us to alter conditions and get what we want, while our unique psychological ability enables us to rise in creative imagination beyond the actual limitations in which we live. We can dream better and sometimes create better, but we can never escape our essentially finite condition. So the Psalmist said,

> When I look at the heavens, the work of thy fingers,
>   the moon and the stars which thou has established;
> what is man that thou art mindful of him,
>   and the son of man that thou dost care for him?
> Yet thou hast made him little less than God,
>   and dost crown him with glory and honor.
> Thou has given him dominion over the works of thy hands;
>   thou has put all things under his feet,
>   all sheep and oxen, and also the beasts of the field,
>   the birds of the air, and the fish of the sea,
> whatever passes along the paths of the sea.[2]

---

[2]Psalms 8: 3–8, R.S.V.

Similar expression can be found around the world and from very ancient times whenever writers sensitive to the human predicament have written. The dilemma arises from the very nature of the universe, beginning with the theoretical polarities of pure radiant energy and absolute zero between whch each entity and the whole universe must find a relative position.

We shall next consider further refinements that arise as novel characteristics in advanced control systems. The most interesting capability that arises is that of learning, that is, the ability to acquire new patterns for interacting with the environment as a result of experience. Most of the activity of living control systems is genetically determined. This is clearly true of some internal controls in all living entities, but it is also true of many patterns of behavior by which entities relate to the environment. The fascinating behavior of animals has been studied at great length and has made us aware of the intricate and complicated patterns that are genetically determined. These include, for example, the complicated migrations of birds and fish, the mating rituals of various animals, and the dance of the honey bees communicating navigation directions. We may be amazed by some of these activities and the abilities behind them, and our scientists are continually probing the secrets that lie behind these skills. Some animals have abilities that utterly mystify us and have led to many strange beliefs, including the idea that some animals are superior to man. There has been a recent rash of this sort of thinking concerning the dolphins. Some animals do have some senses that are superior to ours for certain kinds of functions, and there is no question that some animals have genetically determined abilities we have not and do not even understand.

Thus, of all the advanced capabilities of control systems, the most interesting and important, in terms of the model we are developing, is that of learning. Genetically determined patterns are relatively fixed, and change is not likely within a single generation. This means that, no matter how remarkable they may be with reference to a certain set of circumstances, they can prove completely ineffective when those circumstances change. In the backyard of a previous residence, I had a shack in which I did some writing and which I shared with a large number of wasps. I was impressed again and again that the invention of glass windows had brought about a condition the wasp could not take into account. When the windows were open, the wasps would fly to the screen and land to investigate. When the windows were closed, they would try to fly right through the glass. This is not to discredit wasps since we find it necessary to paint names and designs on glass walls and doors in order to keep humans from making the same mistake. It does illustrate, however, the point that, no matter how excellent a genetic endowment, changing circumstances may prove too much for it to handle. Robert Ardrey in *African Genesis* reports a fascinating investigation into the knot-tying abilities of a small bird. In the process of making a nest, this bird ties a complicated knot. The investigator isolated several generations from the parents, so that there could be no learning, and kept the birds confined and away from both their natural habitat and other birds. After several generations, the birds were released and observed as they again began to mate and build nests, and they again tied the complicated knot. It seems clear that the knot-tying is genetically determined, not learned. If the environment had no material suitable for that kind of knot, that particular genetic pattern would be useless. The bird would either have to learn a new way of building nests or give up nest building. Alteration in genetic endowment usually takes a long time and many generations; thus for some species it may not come soon enough, whereas for others changes in genetic endowment may have proceeded too far in one direction and led to its extinction. Learning, on the other hand, makes possible some adaptation within a particular lifetime. Of course, what is learned is not always good for the organism and may actually be destructive. However, the possiblity of learning means that some organisms will be able to do something creative (novel) about their situation in the direction of more adequate interrelations.

Here it may be well to pause in our progression and define the term "creative." We shall use it as roughly synonomous with "synthetic." This does not represent an arbitrary choice. It is the most generic meaning and is in accord with the nature of the universe as we have come to know it. All advances, from energy to human beings and their societies, have come from the synthesis of existing forms. Any creative activity, short of creation *ex nihilo*, which is attributable only to an omnipotent god, is a process of synthesizing existing elements into new and more complex forms. In this sense, when we speak of creative imagination, we refer to that function of the mind whereby existing elements are synthesized into novel entities. This applies as well to art or science, mathematical theory or dreams. All kinds of learning can be understood as some sort of synthetic process, a "putting together" of new connections, even if it is no more than a new synthesis of a certain neurological pattern and a certain sensory pattern, as in simple conditioning. This putting together to form a new entity is synthesis.

In terms of the model we are using, the universe is characterized by continual creative movement, which has led to the creation of entities capable of creative adaptation to novel situations. Dreams, myths, logic, mathematics, literature, art, and science, are all examples of creative products. Of course, not all creative products are of equal value, and some are destructive. More effective ways to rob banks, commit murder, and torture people are also creative to the extent that they are novel syntheses. Much of the confused thinking in our time is due to the naive assumption that if something is creative it must also be good. However, if this use of the word "creative" is not to run into serious difficulties, what has been said here about the degree of a characteristic must be kept in mind. When an experimental rat learns to run a maze, it is being creative, that is, it is putting together neurological systems and sensory systems to create a new pattern of behavior for a new situation. (If we are not to fall into the trap of anthropomorphism, we must consider this as creativity at a second or third remove as compared with the creative activities of a human mind.) The capacity for learning, then, means the possibility of there being creative entities within a creative universe.

Chapter 6

Psychic Systems

The material in the preceding chapters has been derived from sciences that have been established for a long time and contain a sizable body of commonly accepted information. In this chapter, we come to an area of human knowledge where there are a variety of approaches, for there is no universal agreement on the nature of the system that constitutes the human psyche. At most we can indicate in broad outline the implications of general systems theory for the understanding of psychic systems and the way in which this point of view correlates with some representative and well-supported schools of thought. This chapter will be devoted to the brief description of a few such schools: behavioral psychology, the developmental psychology of Piaget, and some of the general characteristics of depth psychology. Chapter 7 will be devoted to the application of some of the findings of general systems theory to psychic systems in general.

To begin with, it is necessary to say something about the term "psychic systems." Learning takes place in animals to which no one would attribute mind or psyche. This kind of learning, indistinguishable from the simplest levels of learning in humans, is composed of chemical or neuronic systems of relations. In higher animals, and especially in humans, learning becomes involved in psychic systems of a complex order which can attain a fairly high degree of autonomy relative to the whole organismic system. On the other hand, there are no psychic systems so autonomous as to be independent of an organismic system. In all living systems, learning systems appear to be specialized control systems within an organism.

"Psychic phenomena" have been brought forth to suggest the existence of disembodied psychic powers and systems. There is no hard evidence however, to confirm this and, the model of the universe being developed here makes the existence of disembodied or nonbiological psychic systems most unexpected. That is, the whole of nature appears to be built up of layers of systems, with the highest always participating in and dependent upon all of the lower levels for its organization and continued existence. If there were independent psychic systems, it is difficult to see how they could be a manifestation of the basic energy of the universe. All that has been developed in preceding chapters points to a gradual process of differentiation from level of existence to level of existence, and it is only at the latest and highest level of complexity and specialization that intelligence appears possible. Of course, these two arguments do not "prove" that there are no nonbiological, psychic systems; they could exist as manifestations of an "otherworldly" kind of energy of which we have no ordinary knowledge. Even then, it is difficult to understand how such systems would relate to anything in the universe as we now know it.

In any case, this argument makes the choice of the term "psychic" confusing because it is so often associated with "psychic phenomena." The word "mental" is even worse because it is so closely associated with intellectual processes in the human mind as distinguished from emotional or neurological processes. The term "cybernetic" would be very useful in that it relates to varieties of control systems, but the focus of this chapter will be limited to psychological systems in human beings. The term "psychological systems" however, is ambiguous. Are psychological systems systems within a human psyche, or are they systems of psychological theories? Strictly speaking, they are the latter.

However, the problem is not just a matter of terminology. This chapter will consider those organismic systems that are the result of learning. Since the organismic system does not exist on only one level, learning within the organism does not exist on only one level. What is true of the organism appears to be true of its learning systems; that is, differentiated levels of systems are, at the same time, participating subsystems in the whole organismic learning system. In other words, there is no way to separate "mental" from "chemical" or "neuronic" except in terms of specific degrees of relative autonomy. The human organism learns on a variety of levels, and these levels are interrelated in the control systems determining the behavior of the organism on various levels. What is most important here is to develop the best model possible given the present lack of unanimity in the understanding of the human psyche. As there is no sharp line between life and nonlife, so there is no sharp line between organism and mind. When one looks at the extremes, the mind of man and the learning of simple organisms, the differences are obvious and striking. If one moves step by step through the various degrees of learning ability, no sudden break is observed, but rather a gradual transformation. For this reason, and because of a lack of any better terminology, we shall use the unsatisfactory term "psychic system" to refer to the systems that result from the learning processes in human beings.

We have seen that living entities are steady state systems that must be maintained by the constant use of "packaged" solar energy and their survival necessitates control systems that will preserve homeostasis by the relief of a

Psychic Systems

variety of stresses. For animals, this becomes a problem of relation to a wider and ever changing environment. Hence, the possibility of adaptive behavior becomes very important. In very general terms, adaptive behavior involves a change in some of the patterns of the internal relations of an entity which result in alterations of some patterns of external relations thereafter. In higher animals, this usually means that some new neurological connection or association has been established between sensory and motor patterns which becomes an elaboration or modification of the existing system. One of the central problems of psychology is the nature of the system that mediates between sensory and motor patterns. The physiology of the sensory and motor processes is rather well understood and agreed upon, but the physiology and psychology of the complex system between them is still a subject of considerable disagreement and in need of much clarification.

There is increasing evidence that there is some kind of learning even among the most simple forms of animal life.

> The effect of past events in modifying present activity, the essential feature of what we call habit and memory in psychological terms, is also evident in organisms without nervous systems. As Jennings has shown, protozoa are teachable and learn to reject harmful substances after a few trials.[1]

Experiments on all kinds of organisms from the most simple to the most complex are providing us with an increasing body of information concerning the ways in which learning takes place. We cannot, in a single chapter, cover all the many advances that are being made, and we must also keep in mind that there are still some very important questions for which we have no established answers. Fortunately, for our purposes, we can stay within the limits of what is generally accepted as reliable in light of our present state of knowledge. Thus we shall limit our discussion to what is known about learning in the higher animals, specifically human beings.

Chapter 5 discussed the need to think of characteristics as being present in entities in varying degrees. Learning is another such capability: Different entities are capable of different degrees of learning, and the more complex the organism, the greater the learning ability. It is not known whether this can be shown on anything like a step by step progression but, where large differences are concerned, it seems fairly evident. Further, there is some indication that a significant increase in learning ability is accompanied by a decrease in instinctual patterns, particularly in the case of human beings. Learning, at its simplest level, is the acquisition of a new systemic relation between a specific kind of response and a specific kind of stimulus situation. It may involve a new response to a new situation, an old response to a new situation, a new response to an old situation, or an old response to an old situation within a new frame of reference. Learning is an addition to the system of adaptive behavior within the organism and, as suggested in the last chapter, is, therefore, a synthetic product.

---

[1] Sinnott, Edmond, 1950. *Cell and Psyche*. New York: Harper.

If learning were limited to such a simple, one-to-one relationship, the task of the organism would be practically hopeless. There are too many different entities and different situations in the environment. Useful economy in learning is possible because nature exists in types of entities and situations which are, therefore, classifiable. At the level of the cell, the situation is fairly straightforward and clear. Present indications are that atoms and molecules are "recognized" by the cell in terms of their chemical properties, that is, whether or not they will react with the entity that does the receiving. Larger macromolecules appear to be "recognized" in terms of their geometric shape, that is, whether or not they will fit together with the entity that does the receiving. Because atoms and molecules of the same kind are identical, the systems work well. This is not to say, however, that there are no mistakes. Mistakes can happen because, even though individual entities of the same kind are clearly the same, entities of another kind may not be distinguishable from them. Some kinds of poisoning are the results of such mistakes; for example, an enzyme may take the wrong mate, not get free again, and, consequently, be unable to do its work.

At higher levels, there are even greater difficulties, and here we run into another line of progression. As living entities become more complex, there tends to be greater variability possible from individual to individual within a given kind or species. One wasp tends to act like another wasp, but dogs are more likely to vary from one to another, and human behavior is often very difficult to anticipate. The present state of knowledge in the various levels of scientific investigation tends to support these ideas. It is common knowledge that classification in the physical sciences is highly accurate because of the consistency of the structure and behavior from entity to entity. From there on up through more complex living forms, however, classification becomes more difficult, less accurate, and more uncertain. This is in part because the use of classifications must ignore some important differences from individual entity to individual entity. "Judgments" made on the basis of classification or generalization become more and more subject to error as one moves up the scale of complexity. There are endless illustrations throughout the living system that such error can be helpful or harmful, depending on the circumstances. There is a flower that looks so much like a particular variety of female wasp that it is successfully pollinated by the mistaken attentions of the male of the species. There is a kind of moth that is born and lives in discrete groups with color variations among the individuals so arranged that, when they instinctively land together in a set pattern, they can be mistaken for a flower. Although the same kind of mistakes under other circumstances prove fatal, the survival of particular species and the continued evolution of the whole living system provides evidence that either a sufficient number are relatively free from error or else that many errors are relatively harmless or favorable. Generalization and classification are essential aspects of all learning, but it may be helpful to consider some other aspects first and then return to the nature of generalization and classification in that context.

Learning is a way of acquiring patterns of behavior or making existing patterns more exact and adequate. Learning begins with entities that can "record" past experiences and then "recognize" similar experiences as they come. Learning, however, is a complicated subject about which there are still some significant areas of disagreement. We shall, therefore, take the most generally accepted information from three different areas of investigation and relate it, as best we can, to a general understanding of the development of psychic systems in man. We shall examine some of the basic ideas about learning that follow, first, from the models of conditioning and instrumental learning, second, from Jean Piaget's developmental theory, and third, from psychoanalysis and the practice of psychotherapy in general. Despite a growing number of correlations from one of these areas to another, they still remain separate fields of investigation. What follows is therefore an attempt to relate the most reliable and helpful information from each.

Behavioral Psychology

A number of concepts have emerged from the researches of behavioral or experimental psychologists to give us considerable and useful understanding of the kind of learning that appears to be common to most forms of animal life. We shall consider both classical, or respondent, conditioning and instrumental learning (operant conditioning). Although they often use the same or similar concepts, it is useful to treat them separately because in one approach the organism is considered relatively passive, in the other active.

However, two important types of sensitive adaptation can not be considered examples of respondent conditioning. The first is sensitization. Strong stimulation may produce a sensitivity in the organism such that thereafter any strong stimulus may elicit the same response. This may give the appearance of classical conditioning, but it differs in that the response is not associated with a specific new kind of stimulus. Rather it is a generalized response to the strength of any stimulus. It does no more than increase the likelihood of some natural response. The significance of sensitization can readily be seen in reactions of generalized fear after a very threatening experience. The fear response may be activated by any strong or sudden stimulus, whether threatening or not. The second type of sensitive adaptation, somewhat the opposite of sensitization, is habituation, the cessation of a natural response after the stimulus has been repeated too often. This can be a useful adaptation when continued response would be destructive, but it may be maladaptive by making the organism insensitive to elements of the environment at the wrong time. Both sensitization and habituation are important adaptive aspects, but differ significantly from "true" conditioning.

Respondent Conditioning

The analysis of respondent conditioning, on the other hand, clarifies the processes by which specific kinds of objects and events in the environment may become stimuli in relation to certain existing responses. The aspects of the environment that normally stimulate these responses are called unconditioned (unlearned) stimuli, and conditioning occurs through their repeated association with ordinarily nonstimulating aspects of the environment. If an organism repeatedly perceives a given object or event just prior to becoming aware of an unconditioned stimulus, it learns to respond to that object or event in the same way it responds to the natural stimulus. The new object or event is called the conditioned stimulus, and the organism will, in time, respond to it even in the absence of the unconditioned stimulus. The classic example, of course, is Pavlov's dog. The dog salivates in response to the smell of food (the unconditioned stimulus). Repeatedly ringing a bell (the conditioned stimulus) just before and while the dog begins to become aware of the smell of the food conditions the dog to salivate when the bell is rung even if no food is present.

This fairly limited kind of learning may serve to illuminate some aspects of human experience. Complex experiences that heighten anticipation of pleasure or pain may be built up in this way.

However, such conditioning does not necessarily remain after it has been established. It must be reinforced. In respondent conditioning, reinforcement is accomplished by pairing the conditioned stimulus with an unconditioned stimulus. If the two are always paired during the training period, then the organism receives total reinforcement; if they are paired at irregular instances, the organism receives partial reinforcement. Oddly enough, partial reinforcement is the stronger of the two and the kind of reinforcement most often received in the usual conditions of life. Partial reinforcement is important to an understanding of the persistence of feelings and attitudes even in the face of evidence to the contrary. On the one hand, the strength of superstition and belief in magic are enhanced by partial reinforcement. On the other hand, the strength of individuals to stand their ground under great pressure is also enhanced by the same process. If all reinforcement stops, the conditioning will eventually fade. This is known as extinction. Conditioning that is the result of total reinforcement extinguishes faster than the result of partial reinforcement, but, in either case, if reinforcement is renewed, the conditioning will be strengthened. Sometimes, however, extinction is not final even in the absence of reinforcement; that is, the conditioning may return after extinction in spontaneous recovery. It may, in fact, require continued periods of extinction for the loss to become complete.

A specific response is not associated with a specific stimulus. Rather, a specific *kind* of response is associated with a specific *kind* of stimulus. Organisms conditioned in response to a perceptual stimulation respond the same way to any sufficiently similar perceptual stimulation. For example, organisms conditioned to respond to a sound or color continue to respond when the sound or color is changed in small degrees. Eventually the difference will be too great to elicit that response, but there may not be a clear cutoff point. Responses may become weaker as the differences increase and then stop all together. Experimental psychologists have found it possible to develop graphs of generalization gradient that show the strongest responses to be where stimulus similarity is the greatest and to fade in either direction of differentiation.

Generalization is only half the story: the other half is discrimination. Sharper discrimination is learned if the more similar stimuli are reinforced and the less similar are not. Such discrimination is actually taught to laboratory animals: The experimenter reinforces only those varieties of stimuli to which he wants the organism to respond. Interpretation at this simple level of learning is a combination of generalization and discrimination that represents some sort of organismic "judgment" as to which particular stimuli belong to the class which calls for a particular response. Part of the difficulty in considering sensitization and habituation as learning is the absence of significant discrimination. Each, by its very nature, tends to be extremely generalized on the basis of superficial similarities and, consequently, the information value is very low.

Operant Conditioning

We can consider the second major area of investigation, instrumental learning or operant conditioning. This kind of learning or conditioning is instrumental because it is a means of gaining or achieving something and operant because it comes about as a result of operations of the organism on the environment. It differs from respondent conditioning in that it results from a new association between some behavior of the organism and its consequences. As the organism acts on its environment, some acts are reinforced, (they succeed in attaining certain satisfaction). Most learning in the more complex organisms is instrumental.

From birth, children are active (even if the activity is only crying) in relation to the environment. As children grow, society provides them with rewards or punishments, encouragement or disapproval and thus controls the kinds of reinforcement their behavior will receive. Because satisfaction of their basic needs is directly dependent upon parents and society, children must learn how to be acceptable. This is not just a matter of learning how to do things. Children must learn how to see the right way, what kind of attitudes to have, and how to evaluate themselves. Becoming aware of themselves is part of this gradual process of learning. Thus learning for humans is, at least through childhood, largely a matter of socialization. Children do, of course, learn how to survive and satisfy their needs but at the same time, do so in a way that makes them acceptable to their society.

All that we have said concerning generalization and discrimination applies to instrumental learning. Children learn to associate a pattern of behavior with a class of situations and to make their classification more exact as society provides them with selective reinforcement. They also learn to interpret situations by discriminatory classifications in association with the "right" judgments concerning their relative value. Two distinct problems arise in connection with the truth or reliability of this kind of learning. The first is whether individuals learn accurately enough and become sufficiently conditioned to make a successful adjustment to their society. The second is the relation of what individuals are learning to the actual universe, as well as the problem of determining how reliable and accurate the folklore or wisdom of a society really is. Humans are thus confronted with a twofold problem: they cannot survive without learning, and they cannot be sure that what they are learning is reliable.

Human learning and knowing are, by their very nature, processes of generalizing, discriminating, systematizing, interpreting and judging. There is no way to change the functional nature of these processes, but it is possible to greatly improve their reliability and accuracy. We have already observed that reinforcement is somewhat different for instrumental learning in that it comes as the consequence of some action on the part of the organism, but there are other differences. There appears to be more variety possible in the arrangement of partial reinforcements and a comparable variety in rates of extinction. In other words, there are many degrees of conditioning, and there are many degrees of difficulty involved in losing or altering conditioned patterns. Our analysis of operant conditioning brings to light a much clearer understanding of the processes by which societies have been successful in obtaining a fairly high level of conformity from genetically unique individuals and of the extreme difficulty in alleviating neurotic and psychotic problems. Conditioning, then, is both a blessing and a curse, a necessity that may become a serious handicap. Conditioning is essential for the stabilization and economic function of a personality–organismic system, but it can make a person rigid, unable to adapt to change—even nonfunctional, as in neuroses and psychoses. The results depend on the nature and strength of the conditioning and the interrelations of conditioning situations. However, before we can be more explicit about these ideas, we must bring is some more information from the behavioral psychologists.

Experimentation with methods of shaping behavior has been very revealing. What psychologists call the method of successive approximation is something like the childhood game of "hot and cold." One player has to find the object the other has in mind. The first player is given the clues "warm" and "hot" as he gets closer, "cool" and "cold" as he moves farther away. In the method of successive approximations, acts that are closer to the desired behavior are reinforced while those which are farther from the desired behavior are not. Reinforcement is always relative to the new situation, that is, what was closer at position 1 is farther away at position 2. Hence, the specific behavior which is reinforced is based on successive approximations. By this means, organisms can be trained to perform specific kinds of behavior in response to specific kinds of stimulus situations. In the same way, an organism can be trained to discriminate among similar kinds of behavior and to limit its response to a specific, "right" kind. It is not difficult to see how this relates to the maturation of children.

There is another side to the picture that has much to do with the strength of conditioning and the rate of extinction: negative reinforcement. In this case, behavior is reinforced by the cessation of an unpleasant or painful stimulus. Negative reinforcement can produce two different kinds of conditioning: escape conditioning or avoidance conditioning. In escape conditioning, the relations of kinds of reinforcement and rates of extinction are more or less the same as in positive reinforcement but, with avoidance conditioning, something very different occurs. In escape conditioning, the organism learns a pattern of behavior which will bring relief from an undesirable stimulus. In avoidance conditioning, the organism learns a pattern of behavior in response to fear arising from an anticipatory interpretation. Avoidance behavior may be a combination of operant and respondent conditioning, that is, the organism learns to fear certain stimuli which have become associated with a painful stimulus situation. It then responds to the fear and avoids the original stimulus all together. On the one hand, this has great survival value, on the other, it may be limiting or even harmful. Because the organism avoids the original stimulus, there is much less chance for further learning or adaptation in that area. This appears to be the case with many traumatic and painful situations. The organism may go on avoiding where avoidance is no longer necessary, or worse, through erroneous generalization, discrimination, and interpretation, may avoid the wrong situations.

In laboratory situations, avoidance behavior may extinguish suddenly, there may be strong resistance to extinction, or the response may grow stronger even without further reinforcement. This deviation from the other extinction patterns is of great importance in understanding some difficult aspects of human behavior. It is, no doubt, due in part to the organism's shift from responding to the painful stimulus to responding to an anticipatory interpretation so that "reality testing" is blocked. It may also be due in part to a poorly understood, but very common, phenomenon, the fear of fear. In both fear and anxiety reactions, there is a kind of exaggerating resonance that increases through stages of increasing totality. At the first level, one feels a high degree of fear or anxiety. At the second, one becomes afraid or increasingly anxious caused by fear or anxiety, and this brings on a rapid increase in psychophysiological reactions. At the third level is a paralyzing fear or anxiety in response to the loss of control, the feeling of being overwhelmed by fear or anxiety, and the fear of being immobilized in a threatening situation. Fear of the threatening situation is compounded by fear of fear and made total by the fear of loss of control to an overwhelming force. It may well be that these reactions have something to do with a conditioned fear or anxiety response of the conscious mind to the responses of the endocrine and autonomic nervous systems; that is, once fear or anxiety responses become systemic, much of the response pattern is beyond conscious control, and frantic efforts at control may well intensify the total fear response.

Avoidance conditioning plays a significant role in several areas of human behavior. For example, in the case of the method of successive approximation, as it occurs in actual living situations, there would be not only positive reinforcement for movement in the right direction, but also punishment, threat, or pain for movement in the wrong direction. If the pain is sufficient to produce avoidance conditioning relative to certain kinds of behavior, that kind of behavior may be blocked or avoided in general for all stimulus situations. That is, if the conditioning were strong enough, the organism would be blocked from testing that kind of behavior for any other situation. Clearly, this could lead to significant limitation of imaginative play and creativity in problem solving. Further, if the painful or undesirable stimuli were encountered in connection with experimentation, and the conditioning were strong enough, the organism might well avoid all patterns of behavior not well established and socially validated. If we place these suggested possibilities against the tendency of societies to resist deviation and social experimentation, we can begin to explain the high level of conformity within most human societies. These observations may also help to explain the rather common dislike and distrust of those with different social patterns of behavior.

Following this line of reasoning, one can apply an understanding of avoidance behavior to another area of difficulty. Actual situations in life are not as consistent as might be arranged in the laboratory and, consequently, reinforcement and painful stimuli do not consistently result in identical behaviors. In some cases, this leads to increased perseverance, as is the case with those who learn to live with hardship, disapproval, and discouragement. If the painful experiences are too strong, however, the conflict of positive and negative reinforcement can lead to high levels of tension and anxiety, and even withdrawal symptoms. This is not a simple matter because there are so many variables, but it seems clear that an adequate understanding of human behavior requires an understanding of the nature of conditioned learning.

Connections and Relations

According to the model being developed here and our present knowledge of human physiology, learning takes place within the central nervous system and the endocrine system. The biological structures constituting these systems develop largely through genetic control, but the specific connections and interpretations through which sensory input is processed, evaluated, and transformed into neural and motor responses must be learned and developed through activity and interaction with the environment. Behavioral psychology does not take us very far into the complexities of highly organized behavior, including thought processes and self-awareness. It does, however, present us with a clear model of the way in which some connections and systems of relations are formed. Consistent with what we know of human physiological systems, it explains synthetic processes through which novel structures emerge. It does not provide a model for a whole psychic system, but what it does present is not fundamentally different from the universal creative processes described in preceding chapters. As far as they go, the basic ideas of behavioral psychology support a system model of human beings and their universe. Some behaviorists may take a reductionist point of view but, if one ignores their extrapolations and considers only their experimentally verifiable findings, the conflict disappears.

There is more to human psychology and personality than conditioned responses. System and order govern complex interrelations that have great importance in determining behavior, motivation, and creative activity. Probably no behaviorist would disagree with the last sentence, but they are not likely to shed much light on the subject as long as they are committed to analyzing observable behavior and opposed to introspective and clinical evidence. Obviously, there is no way to get at the psychic systems of an individual without depending on introspective reporting and clinical evaluations. These methods may lack the accuracy and reliability of experimental laboratory methods, but they will have to suffice until we develop some highly sophisticated electronic method of brain tapping.

There is an important issue here that must be faced squarely. Is it better to bring as much scientific, experimental, and logical rigor as possible to the best methods available at a given time, or is it better for investigators to limit their activities to areas where they can feel secure in the application of strict experimental method, leaving the rest of life to witchcraft, shamanism, and folklore? In this existential age, it no longer seems justifiable to avoid the hard business of doing the best we can to provide the most useful world and life perspective possible. Granted, our generalizations become more tentative and less verifiable the broader they become, but those of us who believe that reason is a better guide than blind conditioning leading blind impulse have no other choice but to build as well as we can today and rebuild tomorrow whenever we must. In any case, this will be our rationale as we turn from the verifiable evidence of behaviorism to the unsettled areas of developmental psychology and personality theory. There is considerable disagreement and much need for further clarification, but some ideas that have been developed are receving wider and wider support. These perspectives and systems appear to fit well with what information we have and are useful in integrating and explaining humans and their world by a view that then becomes open to criticism and improvement. We shall limit ourselves to two of the many points of view available: Piaget's development theory and some of the more common ideas that are generally acceptable among depth psychologists.

Psyche and System

The psychic functions that guide behavior, including thoughts and feelings, operate with interpretive systems or contexts. In order to move beyond conditioning processes to an understanding of the human psychic system, we need a model that can account for the building of ever more complex structures of information and behaviors. How is it, for example, that children move from stage to stage of increasing mental ability and complex physical skills acquired from their interaction with the environment? Whatever else the psyche may be, it is a control system whose obvious purpose is to guide the behavior of the individual toward survival and satisfaction. It develops over a long period of time and, then, only in interaction with a human society.

A considerable body of opinion and evidence points to the conclusion that the psyche develops and functions in levels, and a number of models are now available to show how these levels function and interact. However, in the present context, not all of these models are equally useful. In light of the preceding delineations of physical and living systems, one is led to assume that the human psychic system has certain characteristics. First, one would expect it to be developmentally continuous with the whole organism. There is no reason, from the evidence so far, to suppose that the mind is something very different from the rest of existence and superimposed on the organism.

That the mind differs essentially from our physical and the biological nature has been the prevailing opinion throughout human history, but that does not tell us anything about the reliability of this point of view. Recent discoveries have discredited many long-standing beliefs. Further, since the whole of existence appears to be composed of layered systems, it is reasonable to suppose that the mind should have a layered organization, beginning with physiological structures and developing through levels of novelty to the highest level of mental functioning. This, in turn, leads to the expectation that, at various levels, there should be subsystems with relative degrees of autonomy acting in some ways as a whole system and in others as dependent subsystems. Finally, one might expect that organizational problems should arise during the process of maturation and the ongoing processes of adaptation requiring the individual psyche to make adjustments between the polar extremes characteristic of all systems. There are models that satisfy these expectations. Although we have as yet no proof of their truth or reliability, one can give a reasonable account of the nature of the human psyche within the context of the model developed here. It is in this connection that we shall examine the developmental view of Jean Piaget.

Before we continue, however, a few points must be made concerning the use of Piaget's system as a foundation for this part of the model. Because his system does not represent a body of established findings in the field of psychology, some reservations must be kept in mind.

1. His findings are not worked out systematically and consistently for all ages and for all concepts used. His emphasis and interest has changed over the years, and some questions have been left unanswered.
2. His methods are not as exact and controllable as more limited kinds of investigations and by their very nature involve the subjective judgment of the examiner to a significant extent.
3. His results have not, as yet, been verified by a very large number of

independent investigators. However, this is now being done by more and more psychologists and the future will, no doubt, bring forth verifications, challenges, and modifications.
4. As a consequence of the foregoing, it must be remembered that Piaget's system is not yet universally accepted as anything more than one interpretation.
5. Piaget has indicated repeatedly that it is not his intention or wish that his findings be accepted as final truths. He tends to view his own publications as progress reports and hopes they will encourage others to pursue similar researches.

One may well ask why we should use a system that is so untested. There are several reasons.

1. Piaget has arrived at a layered view of the development of the human mind that correlates well with a general systems approach. In other words, his is a layered systems model that has much in common with the model developed in the preceding chapters.
2. His findings are useful for understanding some significant problems of man and society in a way that correlates well with what we know about man and the universe in general.
3. His model provides a productive hypothesis for interpreting historically the development of the human mind and the evolution of religion and philosophy in particular.
4. His model provides a new and helpful element in understanding the cultural evolution of man. If, as Piaget suggests, the human mind has evolved through stages, these changes may well be an important cause of significant cultural change.
5. His findings appear to be reasonably consistent with the emerging body of information concerning the psychosocial nature of man. If one steps back from the detail, over which there is some disagreement, it is possible to see Piaget's findings as elaborations of the behaviorists' point of view and as complementary to an analytic point of view.

With these warnings in mind, we turn to Piaget's system as a good example of how the emergence of the human mind can be explained within the context of a general systems view of the universe.

Equilibration

One of the most important concepts characterizing Piaget's thought is equilibration, and because this concept is central to general systems theory, it may be well to begin our treatment of Piaget at this point.[2] Piaget views the

---

[2]For a more complete explanation of Piaget's ideas, see Ginzburg, Herbert, and Opper, Sylvia, 1969. *Piaget's Theory of Intellectual Development*. Englewood Cliffs, New Jersey: Prentice Hall.

development of the human mind as a process of equilibration between the organism and the environment; that is, equilibration is the process of moving from one state of relative balance through disequilibrium and on to another state of relative balance. Development is affected by physiological maturation, experience, and social transmission, and equilibration may be viewed as the process of integrating, organizing, and balancing these effects in relation to the environment. As development continues, the equilibration becomes more fully elaborated in three ways. First, the field of application (objects and properties acted upon) is increased. Second, mobility (the spatial and temporal distance between the person and the objects acted upon) is increased. Third, stability (the capacity for compensation for changes without disturbing the existing psychic system) is increased. One might say that equilibration is the activity of the organism organizing itself in relation to its environment.

Invariant Functions

Another means of clarifying the central role of equilibration is what Piaget refers to as the two invariant functions, organization and adaptation. The human mind evolves by organizing and reorganizing itself as a result of further maturation, experience, and social interaction. Piaget regards the tendency to systematize processes into coherent higher order systems as characteristic of all living entities; that is, organization is an invariant function for all forms of life. The other function he considers equally universal is adaptation, which consists of two complementary processes: assimilation and accommodation, the ways in which the organism elaborates and modifies its organization. Assimilation is the process by which the organism grasps or understands an experience so that it fits into some branch of its existing system. This process, of course, involves some level of interpretation based on generalization and discrimination. Accommodation is the process by which the existing system is modified in order to incorporate the novel aspects of some experience. Assimilation tends to elaborate the existing system, while accommodation tends to reorganize some subsystems of the existing system to make it more inclusive.

This leads directly to another fundamental aspect of Piaget's thought, his emphasis on the activity of the organism. He sees the development of the human mind as the result of the active interaction of the organism and the environment. This may seem trivial at first, but it has significant implications for the interpretation of human beings and other living systems. Its importance lies in the change of perspective involved in the difference between a conception of a passive organism being conditioned by its environment and the conception of an active organism pressing on its environment and acquiring new ways of relating as it creates its own inner organization in the process. The new ways are rather limited until one gets to the level of more complex organisms, but, with the emergence of the human brain, the capacity for variability becomes striking. In fact, the combination of an active assault on the environment and the relative absence of instinctual behavior patterns in people indicates that the development of the human mind and personality is largely the result of dynamic interaction between emerging individuals and their environment.

Sequence of Developmental Stages

The process begins with four aspects of the genetic endowment of the individual: physiological structures, automatic patterns of behavior, the invariant functions, and the necessary sequence of developmental stages. Piaget does not concern himself with the first two except to indicate that they are the foundation for all that follows. By physiological structures, he means the whole organism in general, but especially the sensory, neurological, muscular, and hormonal network which make all higher levels of development possible. By automatic patterns of behavior, he refers to the simplest level of responses, such as reflexes, which require no learning. In his view, these simple responses are relatively insignificant when compared with the impressive hierarchy of acquired abilities. A consideration of the necessary sequence of developmental stages will take us into the main body of Piaget's system.

Piaget describes four stages in the mental maturation of the individual: sensorimotor, preoperational, concrete operational, and formal operational. He does not suggest that these are fully actualized in all persons or all societies. He does insist that the order is necessary and universal. His findings indicate that each stage grows out of the novel attainments of previous stages in the same kind of layered order that appears at all levels of existence. Thus, as individuals interact with their environment, they mature through these stages in ways that approximate the attainments of their society. They consider the order of the stages of maturation to be universal and necessary, although the degree of attainment is culturally relative. An implication here worth emphasizing is that mental maturation, unlike other forms of biological maturation, is not wholly genetically determined. Given proper nutrients and adequate health, an organism will mature as determined by its genetic pattern, but human organisms may be very healthy and live successfully without ever maturing very far relative to their psychic potential. Too close an analogy between physical and mental growth may therefore be misleading. The brain matures on schedule as genetically determined, but the development of its potential depends on its use. In this sense, progress in mental ability appears to be similar to progress in the use of other parts of the body. That is, the muscular and nervous systems can be developed in the direction of many skills, but the acquisition of those skills is not automatically provided for genetically. The skills a particular individual develops are determined by a combination of such things as cultural opportunity, motivation, and practice. This appears to be the case with the development of the human mind as well.

It has a large potential for "mental skills," but the right set of circumstances and practice must be present for these skills to be actualized. To the extent that "mental skills" are developed, they are developed by levels and in proper sequence, entirely consistent with what appears to have happened throughout the whole evolutionary process.

In the first, or sensorimotor, stage, children pass through six substages as they learn to coordinate senses, nerves, and muscles into useful behavioral subsystems.

1. In the first substage, children begin with automatic patterns of behavior and begin to move by means of assimilation and accommodation into the second stage.
2. In the second stage of primary circular reactions, children transform chance patterns of behavior into schemes of learned behavior with emphasis on the mastery of their own body.
3. In the third substage of secondary circular reactions, children are concerned with the external environment and the transformation of chance patterns of behavior into learned schemes for relating to objects around them.
4. During the fourth substage, children tend to coordinate the schemes acquired in substage 3. They learn to make more mobile applications of their schemes (generalize) and to take account of relations of space, time, and causality. That is, they become capable of some anticipation and more complex patterns of imitation. This stage is characterized by a significant increase in the separateness of the object as something to be looked for.
5. In the fifth substage of tertiary circular reactions, children develop an interest in novelty for its own sake. Instead of a mere conservative interest in mastering new schemes, children now search actively for novelty. External objects become more separate as things to be sought out and manipulated.
6. The sixth substage is one of transition, during which children begin to develop mental images and to use words to refer to absent objects.

The model Piaget provides for the first stage sets the pattern for all that follows. Through assimilation and accommodation, separate sensory and separate motor activities are organized into complex schemes, allowing children to expand their field of application, to increase the mobility with which these schemes are applied, and to increase their stability through the development of more adequate schemes, which can be applied to a wider variety of situations and objects. The greater adequacy of the schemes necessitates taking account of spatial, temporal, and causal relations and this, in turn, increases the decentration of the field of experience through gradual objectification. As children learn schemes that can accommodate changes in the spatial and temporal relations of objects and their causal sequences, they gradually separate (dissociate) self from the objects of experience. The development through these substages illustrates the combination of the activity of a child with the need for continual equilibration. Children press for mastery of chance behavior and for greater mastery through sought-out novel experience and seem to be balancing their ablity against the environment of which they are becoming aware.

Piaget presents two principles that bear an interesting and supportive relation to a model of layered organization. The first is the principle of moderate novelty. That is, if its degree of novelty is too little, an object or activity will not hold the interest of the child. If the degree of novelty is too great, the object or activity will not be noticed or "understood" by the child. If the degree of novelty is such that the object or activity is recognizably similar, yet still different enough to be interesting, then developmental learning is much more likely to take place.

The second principle is that the more schemes children have, the more novelty they are capable of recognizing. That is, more novelty will be perceived by a child who has a greater variety of schemes as a basis for recognition. Taken together, these two principles suggest that the more children learn and organize, the more novelty they will recognize and the more they will learn. This fits a layered model well, especially if one considers the image of a tree with expanding and multiplying branches.

However, there is another side to the picture that has not concerned Piaget. There is considerable evidence from casual observation, that an opposing principle is also at work: The more adequate the psychic system, the easier the assimilative interpretations, the less the novelty that must be recognized. This follows, in a sense, from Piaget's definition of stability as the capacity for compensation for changes without disturbing the existing system. This is of positive value for the organism in one sense, and detrimental in another. Individuals have the goal of greater equilibration between themselves and their environment, and this is achieved through the development of a growing system of schemes and operations. However, the greater the adequacy, the greater the likelihood that interpretive assimilation will replace accommodation, and the more degrees of novelty will pass unnoticed. What follows seems to be the principle that the degree of novelty required to be recognized as novelty increases in proportion to the adequacy of the hierarchy. These considerations are not unrelated to a principle derived from the practice of psychotherapy, which states that individuals will move therapeutically only when they experience an optimum amount of pain. If pain is understood as a degree of imbalance, we can say that individuals will equilibrate at a higher level of adequacy only when experiencing an optimum amount of disequilibrium. If the degree of disequilibrium is too great, the system may become nonfunctional. If the degree of imbalance is too little, the system may well ignore it.

This many-sided problem will come up again and again in subsequent chapters. In the present context, where we are dealing with early childhood, active curiosity and accommodation tend to outweigh assimilation because of the need for greater adequacy and stability. However, we shall see later that the tendency to conserve the existing system grows stronger with the development of that system.

The second, or preoperational, stage marks beginning of organized language and is the period of greatest language growth. All we have said concerning the advances of the first stage has been in reference to sensorimotor developments. In the second stage, similar developments begin to move in the direction of the kind of self-conscious awareness we usually associate with the mind. However, the child does not move from one stage to another without going through a gradual process of transformation. Piaget's descripton of these gradual transformations provide us with a most helpful model for understanding many human problems.

Children appear to begin life in an egocentric state in which all kinds of experience are part of one homogeneous, undifferentiated maze of sensations. They are absolute realists, in that they assume that the environment is just as they experience it. In the first stage, children make considerable progress in differentiating the elements of their environment but, for all of their skill, are still very much part of the situation and bound to actions and reactions within their physical context. They have gained enough, in the way of beginning abstraction, to generalize their sensorimotor schemes and apply them to different kinds of situations. As yet, they have no awareness of themselves as separate entities, or of the separation between thought and external things.

It is this very lack of separation that gives rise to one of the chief characteristics of this stage of development. Thoughts can be manipulated at will as in symbolic play and, lacking clear realization of the separation between the actual environment and thoughts, primitive adults probably relate to the world in terms of magical powers. That is, they relate to their world as if some unseen force or influence ran from thing to thing so that the manipulation of one caused the manipulation of another. Children, rather, do not think in terms of forces, powers, or objects; they simply feel and act as though the world were operating the way they feel about it. They have not yet become aware of differences such as arise from perspective, interpretation, and attitude. Hence, during this stage, they are involved in symbolic play and magical explanations.

This lack of clear differentiation appears clearly in the uses of speech that Piaget observed. He found that children in the second stage use speech in two different ways without necessarily being aware of the difference. The first is communication, in which the children are clearly trying to say something to one another. But, since they are unaware of others' having a separate perspective, children tend to talk in a way that feels satisfying without taking into account what is needed to help others understand.

The second use of speech is merely the child's unawareness of others carried one step further. Piaget calls this noncommunicative or egocentric speech because it is speech used for the child's own purposes with no apparent concern for how it might relate to anyone else. First, it includes repetition much like that so characteristic of the first stage. Children repeat speech patterns just as they repeated sensorimotor patterns in the first stage or, said another way, in this stage they repeat the sensorimotor patterns necessary to speech just as they repeated more basic sensorimotor patterns in stage 1. Functional assimilation of a new pattern requires repetition, and so children tend to repeat a pattern until they master or become bored with it. Second, there is what Piaget calls monologue, children talking to themselves. This occurs when a child talks who is alone and engaged in symbolic play. A third example is collective monologue, which differs from the second only in that it involves several children. Piaget observed that, when children of this stage play together, each tends to continue a monologue, simply assuming that the others are sharing in the game.

If Piaget is understood correctly, it is probably misleading to speak of the child's perception of magical powers. He uses the term "participation magic" to refer to the orientation of children to their environment, but he also emphasizes that the terms and abstractions of his system represent the analysis of an adult observer and not reality to the child. Primitive adults are usually aware of a difference between the nonmagical and the magical, even though the latter may pervade large areas of their life. Children, however, only gradually come to distinguish between the "natural" and the "magical." That is, they must experience a series of frustrations and disillusionments before they begin to dissociate or differentiate between inner and outer and between fantasy and reality. This change, too, can be seen in gradual transition.

Piaget has found that children move through a changing emphasis on three slightly different orientations to their environment. The first which has just been described, is participation, by which objects are somehow magically involved in one another. The second is animism, in which view objects are seen as capable of directing their own activity and relations. (An adult would express this as each object possessing its own spirit or psyche.) The third is artificialism, in which objects are viewed as artifacts produced by some humanlike agent. Each of these three orientations represents an interpretation by a child's inner world of the outer world of the environment.

Each incorporates a variety of anthropomorphic interpretation through which the environment is understood as analogous to a child's inner experience. At the same time, each is a further step in the direction of differentiation of the field of experience, so that the anthropomorphic analogy is applied to a more narrow and carefully defined class of objects in the environment. In participation, the humanlike agency is diffuse, vague, and general. In animism, the humanlike agency becomes more discrete and is located in objects as such. In artificialism, many objects are recognized as incapable of selfdetermined activity and are regarded as products of the activities of living or intelligent beings. In other words, as children more clearly define their field of experience and differentiate more clearly among the actual qualities of objects, they become more aware of the differences between their inner world and that world "out there." Awareness of the self as a separate entity, as Piaget suggests, comes gradually through a process of dissociation, a breaking up of the field of experience.

However, this transition is not all accomplished in stage 2. The dominance of the artificialist point of view comes into its own in stage 4 and then appears to become a matter of philosophical and religious clarification and debate throughout adulthood. This transition seems to run parallel to the gradual develoment of intellectual capabilities (see Figure 10).

Much more could be said about Piaget's system and the subsequent advances of developmental psychology. However, the brief description given here may be enough to show that Piaget has developed a system model for understanding the development of human intelligence. He begins with physiological structures that are neuronic subsystems and describes the processes through which levels of more inclusive subsystems (schemes) are developed. This is a synthetic pattern in which new connections form new levels of more inclusive organization, and new subsystems integrate into higher order subsystems. His model also indicates that mapping of self and environment by the human psyche through the development of more accurate generalizations are based on more precise discriminations. In other words, the information system contained in the human psyche comes to map the environment more accurately through successive approximations guided by positive and negative feedback.

Figure 10
Piaget's stages of development.

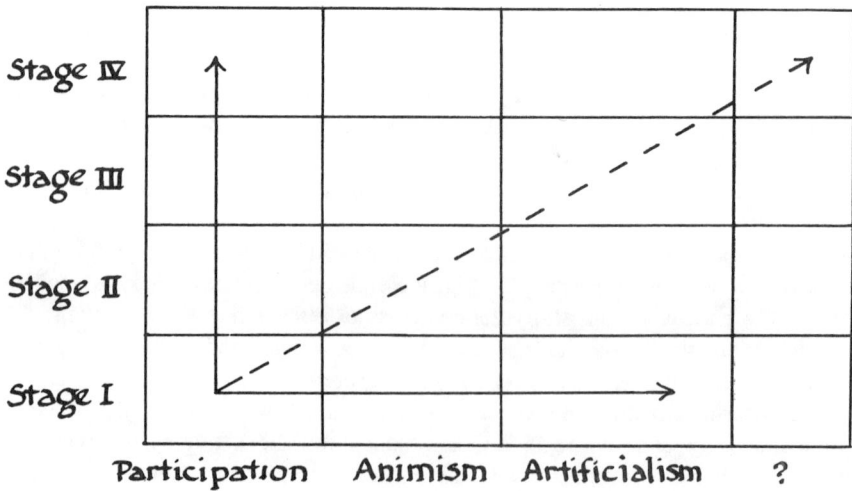

Piaget's patterns of development must be set in a larger frame of reference. The nature of his investigations, as he determined them, required him to set aside considerations of the development of the emotional life and personality although he has referred to them incidentally. It is important to remember that we are dealing here with only one aspect of the total maturation of the individual. Piaget himself does not regard these stages as absolute or total in any sense. A significant part of his findings is his observation of the alternating advances and regressions of children as they made gradual changes. These advances do not necessarily involve all aspects of personality and, owing to compartmentalization and the capacity for regression, a given adult is likely to function on different levels under different sets of circumstances. What Piaget appears to be saying is that the individual is capable of these advances in operational ability at various stages, not that he attains them in any automatic, absolute, or total manner. Seen in this light, his model leaves the door open for other models which are concerned with the maturation of the person as a whole.

Personality and System

We now enter a highly controversial area where there is little certainty about anything. For this reason, we shall make use of only those ideas that are likely to enjoy a fairly large amount of support. Fortunately, the present context allows us to remain rather general and thus avoid the detail over which there is considerable disagreement.

Equilibration

It will be useful to begin with the idea of equilibration because it is so universal and central. In fact, the idea is so largely taken for granted that one is likely to be unaware of how often it appears either implicitly or explicitly in discussions of psychological matters. Psychotherapists often refer to psychological homeostasis, and, in common usage, we often refer to mental health and maturity as mental balance. The individual needs to attain relative balance not only in relation to the environment, and many of those studying personality have been primarily concerned with the problem of equilibration within the individual psyche. If the development of psychological organization proceeded according to a set genetic plan, then one would expect a well-integrated, smoothly working system always to emerge at maturity. However, this is not the case, and the attainment of an intelligent, mature, realistic, and integrated personality is a large problem for which we as yet, do not have very good solutions. It is conceivable that, given the right kind of social situation, the individual might mature as we have described, but no one has yet any very reliable understanding of just what that situation would be. There are many ideas but little agreement and little hard evidence. Some generalizations which have emerged, however, seem to be common to a number of theories.

The process of psychological maturation may be described as a series of stages of equilibration along a developmental line not wholly different from that described by Piaget. This process is often viewed as movement from an original state in which the field of experience is unorganized and undifferentiated to a relatively mature state in which the individual achieves a workable degree of stability. Opinion in psychology and religious intuition have often held the original state to be the source of a deep feeling of wholeness and unity from which later experience removes the individual through the differentiation and specification of elements in succeeding maps of self and world. If the transition from one stage to the next is troubled, the individual may seek a wide variety of substitutes for this lost estate. In other words, the breaking up of the original state through decentration and differentiation must be compensated for by adequate organization and satisfying relations, both internally and externally. Also as in the findings of Piaget, the maturation process is often seen as directly related to the resolution of problems arising from experiences of conflict and frustration. That is, the psychic organization appears to develop through providing solutions to problems of disequilibrium either of an internal nature or between the individual and the environment. There is a reasonable amount of agreement that the largest part of the psychic organization functions at a nonconscious level and consists of various levels of organization (subsystems), corresponding to various levels of the maturation process. Probably, few will question the view that an invariable aspect of the psychic organization is the development of a judgmental subsystem that functions in terms of acquired prohibitions and goals, contains established attitudes toward success and failure, and tends to punish or reward itself with appropriate emotional responses. There is substantial agreement that psychic organization always includes a system of defenses that function to protect the system against serious disturbances from either internal or external sources. Finally, it is reasonably certain that generalized positive and negative feelings about oneself relate to one's degree of successful integration, the adequacy of organizational solutions, the nature and severity of the judgmental aspect, and the successful functioning of the defense mechanisms.

Piaget's description of the intellectual development of children of Western civilization must be seen within this larger context if one is to understand the actual human predicament. The picture would be very different if the whole psyche developed along with the logical and intellectual functions of the human mind. In reality, however, these functions represent a small part of the psychic organization, arrive late on the scene, and are, for the most part, unaware of the workings of the larger and more powerful parts of the psychic organization. Even so, psychological problems would be greatly reduced if there were always an integrative correlation from level to level. In actuality, unresolved conflicts are often walled off and left behind, inadequate solutions disguised, and failures rationalized in order to keep them out of consciousness. Equilibration is often gained and maintained through self-deception and distorted interpretations. In other words, assimilation and accommodation in the process of producing equilibration through the elaboration of psychic organization do not always function in terms of reality. They often enhance the self-judgment of the psyche at the expense of truth. If one thinks back to Piaget's description of symbolic play in preoperational children, it appears quite natural for children to resolve problems with fantasy solutions. One great advantage of such solutions is that they can be as self-enhancing as the child wants them to be, although they can equally be painful and destructive, involving guilt and negative self-judgment. Another advantage is that this kind of solution comes much more easily than those that attempt to deal with the multiplicity of variables in actual situations. Because fantasy solutions are in terms of participation, magic, animism, and the like, they easily overcome many troublesome limitations of actual, finite situations. As long as the individual is reasonably successful and enjoys sufficient social validation, processes of partial reinforcement supported by rationalization and other defense mechanisms keep many unrealistic solutions intact.

A glance back at the history of Western civilization gives an interesting perspective on this problem. Western civilization, like all other civilizations, began with mythological solutions to the problems of man in relating to his universe. These solutions, like those of children, operated in terms of participation, magic, animism, and artificialism and were successful for long periods of time. However, in ancient Greece, Western man moved far in the direction of abstract, objective, formal operations with the development of mathematics and logic. Not until after the Renaissance were the same rigorous method and level of objectivity applied to the understanding of the physical universe. By the nineteenth century, the same approaches were making significant transformations in the understanding of the nature of living things but, not until contemporary times was this level of understanding systematically applied to the self and society. Progress in mental maturation has been fragmentary and uneven, and remains so today. Since the maturation of the individual is culturally relative, it is not surprising that mythological and fantasy solutions have never been totally replaced.

What has just been written suggests that people would develop into mature, rationally ordered individuals if they had sufficiently advanced knowledge and understanding, but this is not necessarily the case. There are other reasons for failure to attain a high level of mature integration. Part of the difficulty lies in the inaccessibility of much of the psychic system. The processes of conditioning, the development of sensorimotor schemes, the elaboration of symbolic meanings and associations, and the recording of an endless number of interpretations from an equal number of experiences all contribute to the building up of a complex psychic system long before anything like self-conscious awareness comes onto the scene. Most of this system is inaccessible to consciousness and much of it is deeply reinforced by heavy emotional charges and/or by its significance within some patterns of the basic system.

In Piaget's timetable, by the time children are able to think in a genuinely abstract and objective manner, they are entering adolescence with a very elaborate and complex psychic system. To suppose that they could by an act of conscious choice at that point become rational and objective is to fail to understand that the psyche is a system and that its various levels and subsystems are only partially integrated and have varying degrees of autonomy. Were we fortunate enough to reach adolescence with a well-integrated psyche based on realistic solutions to the problems of self-evaluation and of relating to the environment, our rational development might evolve and remain relatively congruent with the rest of the psychic system. Most of us, however, are not so fortunate. We arrive at adulthood with a patched-up psyche that manages on top of compartmentalizations, conflicts, and all sorts of unfinished business as long as our environment is sufficiently supportive to keep us afloat. After some necessary amount of what is called adolescence, most individuals arrive at what Freud called a stabilized personality, not necessarily a mature or well-integrated personality, but a workable arrangement of checks and balances, adaptations and defenses so that, in a reasonably cooperative environment, we can function adequately. Psychotherapy has taught us that bringing those deep reaches of the psyche into integrative congruence with conscious rational processes is a very difficult and sometimes seemingly unattainable goal. One of the mysteries of the human psyche is the tenacity with which it hangs on to and defends those parts of itself which are clearly unsuccessful and even destructive.

Approaching the problem from a slightly different orientation let us repeat that the amount of genetic determination tends to decrease, and the need for learning to increase, with more complex organisms, most noticeably in the case of human beings. From this point of view, the brain and nervous system may be seen as highly developed tools for more successful adaptation. Their success in terms of survival and domination of the environment can hardly be questioned. The highest level of human attainment begins at Piaget's stage 4, during adolescence, when the individual develops the capacity for abstract rational processes capable of considering multiple solutions, anticipating consequences, and devising tests of adequacy and reliability. By this time, however, the psyche has done most of its basic learning through trial-and-error probings charged with heavy emotional involvement and subjected to the rewards and punishments of conditioning forces. Our central source of difficulty is that, long before we are capable of critical rationality, we must learn how to behave, to evaluate ourselves, to interpret success, and to make our various psychic potentials conform to the patterns we are learning. Because this learning is done at an emotional or affective level, an intimate connection evolves between our emotional development and our capacity for self-evaluation. Further, because this learning must be done within the interpretations of a cultural context, an intimate connection evolves between self-evaluation and the problem of adjusting to preestablished cultural standards.

The problem is compounded by the fact that we are always potentially deviant (constrained variety) and must be strongly conditioned long before we have any understanding of ourselves or our situation. The possibilities of deviation become manifold through the imaginative reconstructions of symbolic play and become more refined and sophisticated at higher levels of mental maturation. Children actively pursue satisfactions, not some abstract notion of truth, and the sought-after satisfactions are many and conflicting. Because conditioning processes are rarely consistent or constant, the resulting partial reinforcements are likely to be fragmentary, disorganized, and accompanied by feelings of taboo, guilt, anxiety, and the threat of alienation.

When these various notions about the human psyche are studied within the present context of systems analysis, the model that emerges is one of the psychic system as a system of systems with one's sense of identity directly related to the integrity of that system. The psyche develops subassemblies of sensorimotor schemes, symbolic association patterns, representational maps, rational constructions, and habitual behavior patterns, which are interrelated into many different parallel, overlapping, and fragmented hierarchies. The more successful patterns of development result in a highly integrated, cooperative, and congruent system of systems that tends to be self-accepting and has a reasonably realistic self-evaluation accompanied by feelings of personal worth and adequacy. The less successful patterns of development result in various degrees of failure in the systemic integration of some of the subsystems. Lack of functional integration results in the necessity for repressive controls against the self-assertive autonomy of subsystems. The rigidity of personality, which seems to increase with either the level of emotional disturbance or the intensity of authoritarian control, may be seen as directly related to the number of nonintegrated subsystems and the fear that any change may set them free.

In fact, it is possible to relate most of the major kinds of mental problems to this sort of model. A failure of hierarchic integration could account for the fragmentation of common varieties of schizophrenia, and the more dramatic cases of multiple personality can be represented as a lack of integration among major subassemblies. In the therapy of such individuals, a new and different identity seems to emerge that replaces the identities which existed in the subassemblies. This is consistent with the notion that the identity of the self is the system because, as the self-assertive energies of the subsystems are drawn into the integrative synthesis of the psychic supersystem, one would expect that the novelty of the new synthesis would produce a novel sense of identity. In a similar way, other types of mental problems can be related to this point of view in terms of the relative success and strength of the psychic supersystem (system dominance) in balance with the self-assertive autonomy of the many subsystems that constitute the whole psyche.

However, most important in the present context is the point that the self is a system and that such human problems as personal meaning, worth, and identity are all directly related to the developmental elaboration and integrity of that system. The self is inevitably a subsystem with a larger social or cultural system and, if it is to be adequately understood, it must be seen in this larger context.

Chapter 7

General Systems Theory and Psychic Systems

As suggested at the beginning of the previous chapter, there is little in the way of a clearly defined and commonly accepted body of knowledge concerning human personality, what we commonly refer to as the self. It would seem strange to find in a biology text a series of chapters comparing 15 or 20 speculative views on the nature of cells. The more complex the system under investigation becomes, the more difficult it becomes to work with direct evidence, and the more unmanageable become the number of variables. Perhaps the most difficult of all areas of human research, at the present time, is that seeking understanding of what is most familiar and intimately related to each of us, that is, of course, the self that is attempting the understanding. We are all familiar with and constantly use terms such as "self," "person," "soul," "spirit," "character," or "personality." We use these and related terms to refer to that in ourselves and others that is the true essence of a human being, that which distinguishes humans from animals. In earlier times, human beings did not distinguish themselves clearly from animals or plants. Often these were identified as having sacred or magical powers, some were worshiped, and some were used to express the identity of a person, clan, or tribe.

However, as self-conscious awareness increased, especially in civilized societies, human beings came to recognize that we are somehow different. More often than not, this led to a conception of human beings as having a spiritual essence or soul inside the body.[1] This model or image was highly developed in ancient Greek philosophy, particularly in the writings of Plato and Aristotle. One's inner essence was closely identified with the rational

---

[1]Allport, Gordon, 1955. *Becoming*. New Haven: Yale University Press.

functions of the mind. It was clear to these writers that humans share the need for nutrition with plants and animals and sensitive capabilities with animals but alone have the capacity for reason, which therefore is our essential nature. Plato thought of the soul (reason or mind) as imprisoned in the body and longing for a return to the eternal realms in which it preexisted. In India, the idea of the inner self, or atman, was seen as the inner essence of the whole universe, the ultimate Atman or Self. In the West, Judaism, Christianity, and Islam followed a similar model but with much stronger emphasis on the integrity of a Person or Self as the source and sustainer of the universe. East or West, primitive or civilized, most humans have believed that an entity (spirit) exists within themselves different from the body and believed that the essential power or powers of the universe are similar in nature to that entity.

As ethical theory and legal systems further evolved over the centuries, it became universally believed that this *inner entity* is the true identity of a person and responsible for choices and actions. The justification for holding a person responsible lay in the conviction that this inner entity is free to choose right or wrong and, if normal, has the responsibility of knowing the difference.

Of course, there have been many variations on the central theme throughout human history. These variations have been for the most part, beliefs in the existence of some of a large variety of entities similar to our inner nature but not hindered by a body. These noncorporeal entities are usually supposed to have unusual powers, including the power to take possession of a person's inner essence and, in effect, change the individual's identity to some degree. If these entities do not totally take over the individual's inner essence, then they may effect thought processes so as to cause a wrong choice, rebellion against what the individual knows is right, or, in the most dramatic cases, delusions of grandeur and action as though the individual were not subject to the same constraints as other humans. Many plays and stories have been written about such tragic figures, usually princes, warriors, or kings.

The purpose of this chapter is to pull together some of the central ideas, presented in Chapter 6, representative of the way modern psychologists approach an understanding of this inner essence of man, and to show how general systems theory is applicable to them. This development is intended to indicate how the general systems perspective and model can accept and make sense out of ancient folk wisdom from around the world. Others, such as Carl Jung, have presented elaborate and useful psychological theories that have been able to correlate in fascinating ways with man's ancient and symbolic wisdom. There will be no attempt here to create a new psychological system, least of all to try to emulate Jung's brillliant work of a creative lifetime. Our purpose is much more simple, to translate psychological knowledge or theory into the vocabulary of a different model in order to show that the model applies to this level of entity as well as it has to those already considered. Among advantages of this translation is, first, that the use of a relatively neutral language can assist in relieving the confusion of overtones and implications that haunt more traditional vocabularies. Second, the fact that general systems terms represent characteristics that appear to be universal for all systems indicates, in advance, a general model for testing and research in every area of human knowledge. Third, the fact that novel characteristics are seen to appear at every new level of complexity suggests that, on approaching a new level of investigation, such as psychic systems, one should look for and seek to define clearly such novel characteristics as do emerge. Fourth, these novel characteristics generally lead from the level under consideration to the next level of complexity, if it exists. However, they do not make the next level predictable because of the wide variety of possibilities that exist at any level and increase with greater diversity.

In this connection, one would expect novel characteristics to emerge at the level of psychic systems (personality) that would make the next level of social systems possible. The chronological implications of this manner of speaking are misleading. In Chapter 2, it was suggested that the primacy of pure energy over subatomic particles was logical or schematic rather than chronological. Similarly, we must now point out that people or psychic systems have never existed without social systems. We know that social systems exist among primates and, that therefore, there is parallel development of higher levels of psychic systems with higher levels of social systems. Nonetheless, it is still consistent to say that the novel characteristics of emerging psychic systems are what constitute and make possible the bonding and internal relations of new levels or types of social systems. It is equally consistent to say that the emerging or evolving social systems make possible, through differentiation, specialization, growth, and other processes, the emergence of new psychic development. (See the diagrams of interacting polarities in Chapter 9.)

Fifth, the model that emerges from these characteristics provides the foundation for the easy transfer and comparison of information from one area of knowledge to another. Although general systems theory does not necessarily add information to any field of investigation, it does provide a model for the organization of existing information and a perspective that may well encourage the discovery and clarification of further information. The theory may also serve as a guide to where such information might be found.

The previous chapter attempted to summarize various investigations into human psychic systems. Now we shall attempt to show that, from the perspective of general systems theory, large areas of agreement already exist that become apparent upon translation of professional terms into general systems concepts.

A further caution is necessary before we come to grips with the detail of human psychology. At every level of investigation, it is necessary to remind ourselves of the polar problems of reductions, on the one hand, and ghost words on the other. There has been a long and fruitless debate over whether an entity is simply the sum of its constituent parts (a reductionist claim), or whether the whole entity possesses some higher unanalyzable nature (an idealist claim). General systems thinking rejects both extremes. Novel characteristics emerge at each new level of complexity that are or are not reducible to the sum of the constituent subsystems, depending on what one means by "sum." If the word "sum" is understood to mean organized into an integrated system of some sort, then an entity is the sum of its parts. The critical point is the bonding of the subsystems into a new system having, as a whole, characteristics differ from those of any of the subsystems. These novel characteristics are not mysterious essences. New levels of internal organization involving differentiation, specialization, and modification of patterns of internal development give rise to new capabilities in terms of external relations.

It may be well to recall, at this point, the initial hypothesis that an entity consists of subsystems, internal relations and processes, external relations and processes, and nothing more. This hypothesis was easier to accept while we were concerned with atoms, molecules, cells, and even plants. However, the closer we move to human consciousness or personality, the more the subtle notion creeps in that there is something more. The only "more" there can be is the wholeness of the system, a wholeness describable only in terms of the system's constitutent subsystems, its internal relations and processes, and its external relations and processes. These are the characteristics of the system, and there is nothing more to say about it. This hypothesis is of critical importance and, if it does not hold, then the model developed here is in serious trouble.

Developmental Continuum

Our knowledge of mammalian growth indicates that there is a gradual development from the original integration of the reproductive cells through the fully matured organism. This development follows the pattern of growth, differentiation, specialization, and equilibration through many stages until, in the mature animal, there are many interlocking complex systems. The intimate relations among inner maintenance and distribution systems that make possible the growth and equilibration of structural systems at every level of complexity is likely to be our best source of reliable and easily obtainable information concerning the potential for hierarchies of systems and parallel hierarchies of control systems. The advantage in studying the animal as a system of systems is that its growth and development follow genetically determined patterns, so that processes and development can be observed at each stage. This allows for reliable generalizations of the biological sciences that in turn, provide the frame of reference or paradigm against which functional variations and dysfunctional deviations can be described.

The description of psychic systems must begin with the development of the electrical and chemical potentialities of the somatic structures that are present in all higher animals, particularly in humans. These somatic systems, from nerve ganglia to the prefrontal lobe, from adrenal glands to semicircular canals, are the beginning of any psychic system of which we have direct, verifiable knowledge at the present time.

The ancient mind–body problem must be raised at this point to show how it appears from the perspective of a general systems model. As the body develops, the organs, tissues, tubes, neurological electrical connectors, and so on acquire their structural form. When they are sufficiently mature, they begin to respond to stimuli, send information, and gradually establish new connections and control patterns through trial and error, feedback, and reinforcement. With respect to these processes, there are no essential differences among the three positions described in Chapter 6. Piaget describes the process as beginning with what he calls automatic patterns of behavior, the early movements of an infant or embryo, which are still the object of intense study. The general systems model suggests that these early movements are responses to stress at some level of the system. They may also be responses to an excess of energy-producing materials at the level of cells or tissues, so that action is called for by the control systems of the cells and tissues. If so, the movement of some part of the body would be meaningless from the point of view of the whole organism and would not involve anything like conscious intent. If the movement succeeds in relieving stress, it is "reinforced" and will be repeated under similar stress. Clearly, some actions are genetically determined and, one hopes, the time will come when their clear delineation will make possible a clearer understanding of the development of higher level control networks. For our present concerns, however, it is sufficient to say (with the behaviorists) that some "unconditioned" responses are present at the level of cells, tissues, organs, organ systems, and neurological systems and some of their interrelations. What is most lacking in development at birth are some of the higher elements of the control systems that correlate some of the relations of the organism to the external world. The systems responding to temperature regulations, stress in the form of pain, responses in the form of nursing movements, crying, random motor activity, and the like appear to be intact at birth. It is clear from the observation of animal births that much more than this level of development is possible. We watch the birth of wild animals on television and see them stumble, walk, and then run within very short periods of time. This is not the case with human infants.

Piaget's description of how schemes or patterns of behavior are built up from existing behavior as well as the work of the behaviorists, provide us with a more analytic view of how and why these connections are made. What is important in the present context is that both describe the development of connections and patterns in a control system. The development of these patterns includes the programming of control centers such as nerve ganglia and brain areas. Piaget's description allows for the build-up of local control or behavior patterns at the level of simple sensory nerves, nerve ganglia, and motor responses into higher order systems. These systems integrate many simple systems into more complex patterns that gradually relate, through cerebral neural connections, to responses of the whole organism to environmental situations. Piaget suggests that automatic movements occur in the infant, which are then repeated until mastered. These automatic movements are already complex from the point of view of the simplest stimulus response levels of control systems in cells and tissues. However, we cannot, at present, state how much of the systematic development from simple cell and tissue responses to larger areas of coordination of overlapping hierarchies is the result of some genetic determination. In some higher animals, as just mentioned, the genetic determination includes complex motor controls and behaviors. However, the general systems model deals with response to stress, and, in the situation we are considering, stress may come in the form of genetically determined chemical or electrical information from control centers, from local or internal stresses, from information arising from stresses in some other part of the organic system, or from perceptions of the environment. Considerable research will be required before we have anything like specific answers. Most likely, whatever genetic determination there may be for sensory motor development simply fades into other motivations, such as the release of local tensions, the sheer enjoyment of coordinated movement, or a striving toward something in the environment.

Given our present state of knowledge, the basic ideas of operant conditioning, combined with a system-building model, could account for the gradual emergence of complex behavior patterns through the gradual synthesis of simple local reaction patterns. The basic model is that systems emerge from the synthesis of existing systems into a larger whole. In the present case, cell responses are synthesized into tissue responses, muscle responses, coordinated muscle response of the whole body, and so forth. The motivation for all this development, as for all systems, can be generalized as stress. The specific work of psychology becomes the analysis and description of the kinds of stresses and the systems of responses that emerge in relation to them.

The advantage of the system model is that it leads one to seek out and analyze the molar responses that emerge when a new level of synthesis produces a new level of wholeness. For example, muscular response is more than (i.e., has novel characteristics, peculiar to itself) the responses of its component tissues or their component cells or their component cellular parts. Similarly, the coordinated movement of a whole body requires the gradual development of more inclusive systems involving the coordinated movement of many muscles in relation to constant feedback and control operations. From the perspective of a human cell, walking appears as a complex system of galactic proportions.

Piaget's description of the first, or sensorimotor, state indicates that sensory and motor learning and their coordination is the foundation for all further development of the human psychic system with its capacity for symbolic and abstract functions. Piaget's work has been subject to considerable criticism, as explained in Chapter 6, but his basic model is a systems model and is consistent with the basic concepts of general systems theory. Moreover, the main outline of this thinking ought to be close to the reality of psychic development and, at the very least, a useful step in the direction of further understanding. His description of early behavior indicates that the psychic system grows with hierarchically organized schemes (systems). The implication is that schemes are related to classes of perceptions of environmental objects. Going beyond Piaget, we should anticipate hierarchies of perceived objects that would relate directly as stimuli to hierarchies of organized schemes (systems and behavior).

The process of learning, if we now combine elements from all three approaches described in Chapter 6, appears to be a refinement of the perception of objects and the development of responses (systems of behavior patterns). We have a clear picture from the experimental psychologists of this process, through their success in training animals to make the more selective discriminations necessary to appropriate behavior.

It seems to be clear from various investigations that learning is basically a matter of acquiring a system of generalizations (symbols representing interpretive perceptions) in relation to a system of responses or behaviors and then continuing to refine both the selective perception and selective behavior. This general description is consistent with Piaget, the behaviorists, and the psychoanalytic emphasis on the need for strengthening the rational ego in its role of pacifying the less selective, less discriminating perceptions and demands of both id and superego.

The processes described here are carried on in laboratories all over the world with many animals of different levels of intelligence and appear to be basically the same in humans at fundamental levels of learning. However, early in the life of a child differences arise that appear to be the novel characteristics of a more complex system. The system is the brain, and the novel characteristics are commonly referred to as the capacity for abstract thought and self-conscious awareness. The human brain is a large system, in both size and complexity. We know that the prefrontal lobe is fundamental to the capacity for abstract thought. Studies with brain-injured patients and experiments with prefrontal lobotomy clearly demonstrate that, without this part of the brain system, there is a serious reduction or absence of abstract thought. The findings of Goldstein, in his work with brain-injured patients, made it clear that those who suffer injuries to the prefrontal lobe are unable to manipulate ideas or symbols of things that are not present to the senses.[2] For example, a patient deliberately let off the elevator on the wrong floor was

---

[2]Goldstein, Kurt, 1951. *Human Nature*. Cambridge, Massachusetts: Harvard University Press.

unable to imagine a floor above or below and think of a means of getting to the right floor. There may be some similarity between these findings and those of Piaget in connection with children whose brains are not yet fully developed. Children require a number of years of growth and development before they can manipulate symbols in their mind to solve problems in the abstract. They pass from an inability to understand the problem through a stage in which the problem can be solved with the aid of objects to manipulate, and finally arrive at a stage where the manipulation can be carried out through the use of symbols in the mind, that is, abstractly. What happens in certain brain injuries or prefrontal lobotomy is the loss of this capacity and with it the loss of anxiety, which is concern for something not present and possibly nonexisent.

It is this capacity of the psychic system to develop systems symbolizing elements of experience and to relate them to systems of responses that develops the capacity to become aware of the symbols as such and to compare them critically with the reality they are supposed to represent. The highest level of objective maturity is usually conceived of as the ability to detach oneself and critically evaluate one's symbols, interpretations, and responses against reality situations under a variety of circumstances. All this seems to suggest that the prefrontal lobe, probably in correlation with other parts of the brain, has a reflexive capacity for monitoring or observing its own functions. It is thus possible for the human psychic system to anticipate situations that might arise and test, through the manipulation of symbols, which of several responses is most likely to succeed. The ability to have such imaginary experiences depends on the programming of the mind through actual experiences that equip the mind with systems of symbols, their relations, responses, and consequences. Abstracted from the system of relations in which they were acquired, these elements can be rearranged to form novel or creative systems of relations. When elements are bonded into new wholes, they generally give rise to novel characteristics, which we usually refer to as invention, creative art, or fantasy. What is commonly understood as the scientific method is a process for clearly stating or defining some symbolic representation of experience and then testing its reliability or probability under carefully controlled conditions.

Value Judgments

In addition to symbolic representation of elements of experience, there is an ongoing evaluation process with reference to the characteristics of the object or situation being represented. That is, part of the representation is information as to whether the object or situation is to be approached, encouraged or avoided. In fact, the symbolic representation, whether images or words, is an interpretation in itself and probably an evaluation at the same time. Given our present state of knowledge, we must assume that most, if not all, evaluations ased on previous experience and that such evaluations are implicit in of the words and images that we use in ordinary conversation. Much of fort of various liberation movements is directed toward bringing to our ess the number of implied judgments contained in our day-to-day and attitudes. It must be assumed that many of these evaluations begin arly in the life of an individual, before there is any clearly understood ess of self. Feelings of fear, shock, revulsion, disgust, nausea, warmth, safety, euphoria, contentment, ecstasy, and the like begin early in primitive responses and, when associated with the image representing an object or a situation, become value judgments. Long before there are words or anything like a conscious moral choice, evaluations and choices are being made on the basis of feelings or sensations in the nervous system and the viscera. What we have learned about systems in general leads us to expect that rational ethical systems and rational moral judgments are novel products emerging from the synthesis of primitive sensations and emotions with later linguistic and logical developments. For example, the ethical ideal of equality of individuals may be a synthetic rational compromise between the primitive desire to be all and the primitive fear of being nothing or totally vulnerable and may be associated with feelings arising out of play and friendship situations. It requires a high level of maturity, or synoptic perspective, to see that the all and nothing extremes of emotional reactions need to be replaced with a more objective analysis of the situation and a more realistic and appropriate decision concerning possible responses. We are thus led by the general systems model to suppose that rational thought processes result from the synthesis of some content of the psychic system into larger relational systems that have the novel characteristics of a new and wider interpretive system. This appears to be the process described by Piaget, a process limited on the one hand by the physiological development of the brain and on the other by the level of learning available in the individual culture: Psychic systems develop in relation to the physiological and social systems in which they participate.

The Nature of Subjective Experience

The central problem concerning human psychic systems is the nature of conscious subjective experience, particularly that aspect of it usually called self-conscious awareness. Piaget sheds some light in describing the gradual development of the ability to think abstractly; recent studies concerning right and left thinking give us a bit more. These studies and experiments indicate that there is a distinct difference in the processes of the two different hemispheres of the brain.

In most humans, the right side of the brain is "nonconscious" of language, logical relations, and linear thinking. It appears to be organized in a more diffuse fashion than the left side and is most successful in analyzing spatial relations. It tends to see things together, rather than sequentially, to respond "intelligently" to stimuli, and to make choices based on a learned set of values, preferences, or probable responses. However, it cannot respond in language and is not aware of itself. The right hemisphere appears to be the seat of creative processes and of the perceptions comprising artistic talent, spatial judgment, "intuitive" judgments, and some aspects of evaluation processes, but without what we know as "self-conscious" experience. It is probably very close to the psychic systems of the higher animals. When experimenters involved the right hemisphere in the perception of an object and asked for a response, they observed that it recognized the objects, made a selection, and responded accurately and without delay. "The conscious person," that is, the left hemisphere, the seat of language and logic, was unaware of the whole process. Half of the brain was perceiving, judging, and responding, but the conscious, or self-conscious, functions of the brain were totally unaware of any aspect of the process. In a normal brain, this unawareness is overcome by the *corpus collosum*, a network of nerve fibers connecting the two halves of the brain. The *corpus collosum* is, at the least, a communications system and may participate in some wider systems functions.

Each half or hemisphere of the cerebral cortex is a specialized subsystem whose function is control of certain kinds of organismic behavior. Each side can make discriminating selections in the interpretation of information, the selection of responses, and the implementation of some responses. For example, in the normal brain, the communications network of the *corpus collosum* incorporates the two complementary subsystems into a single control subsystem but, when the two hemispheres have been separated through brain surgery, they can carry on and develop different standards for evaluating behavior. When the communicative network of the *corpus collosum* is no longer working, each half of the cortex carries on without being aware of the activities of the other. (Interestingly, individuals who have had such an operation to relieve extreme epileptic seizures appear to live perfectly normal lives using whichever control system is best suited to the circumstances.) Careful laboratory testing reveals some interesting limitations and indicates that the psychic system compensates for the missing communications system but not completely. A right-handed person, after the operation, can draw but not write with his left hand and write but not draw with his right hand. The left hemisphere perceives the four corners of a square, for example, as being lined up in a row because its system perceives in linear, logical fashion suited to language. The right brain, suited to handle the spatial structure of the square, fails with the linear process involved in writing, that is, the sequential formation of letters.

These new studies are a further indication that there is a close correlation between the physiological and so-called "mental" processes. In fact, some researchers and experimenters believe that an actual difference in the arrangement of the neurons accounts for the difference in modes of perception and evaluation in the two sides of the brain. In any case, these experiments underscore what was already becoming clear from other sources, that mind is a function of a psychic system that is a neurological system. What we experience as self-conscious awareness is a novel characteristic of a highly developed psychic subsystem. One would not expect a computer to function properly without all of its memory cells, relays, and communication lines present and in working order. It is just as unthinkable from the perspective of general systems theory to expect a mind to function properly without all its neurons present and working properly. It is easy enough to alter the mind and personality with a tiny bit of electricity or a chemical. A prefrontal lobotomy severing the prefrontal lobe from the rest of the brain can put an end to anxiety, but it also may destroy the possibility of abstract thought. Similarly, a few seconds too long without oxygen will cause some of the neurons in the brain to disintegrate and significantly alter an individual's personality. Brain damage from oxygen deficiency often leaves a person so devoid of abstract thought that he responds only to what is present to the senses, and even then without emotion or concern. The individual's characteristic expressions become a flat mechanical response. All the information we have received from studies ranging in scope from psychology to neurology appears to support a model of both the human psyche and the whole universe: New, more complex, or higher levels of the synthesis of existing systems into larger systems produce novel characteristics that in turn, make possible syntheses that were not possible before.

Problems with Words

The most difficult question concerning psychic systems, as mentioned above, is the nature of subjective self-conscious awareness. The combination of a general systems approach, the available information, and some of the speculative insights of psychologists suggests a direction for the clearer definition of the problem. If learning is the more careful definition or differentiation among generalizations (words, responses, and interpretations) by the more careful description and explanation of the objects and situations to which they apply, then rational and scientific solutions to problems must begin by more clearly defining and explaining the nature of the problem. The use of a systems model in explaining the nature of a problem often reveals that there is no problem at all. Put the other way around, the explanation of the problem often reveals that it is not in the nature of the systems under consideration, but only in the interrelations of the generalizations used to represent those systems. More simply, many problems disappear with more accurate definitions of terms. Although it is often the case that the paradigm, model, or other context within which generalizations have meaning may block any movement toward clearer definition, a new model, paradigm, or context may make a clearer definition suddenly seem simple and obvious.

This suggests that a significant part of the problem in defining or explaining the nature of subjective self-conscious awareness, and its relation to the whole psychic system, including the physiological system out of which it arises, lies in the generalizations that have been applied to it throughout human history and the interpretive model they imply. Terms like "soul," "mind," "psyche," "personality," "body," and "flesh" are central villains in the confusion. Coming out of a long and uncritical past, they cover large and complex areas of the human organism and its experience with relatively little differentiation or selection. Current information makes it possible to identify and differentiate many different subsystems and system levels within the vague areas to which these terms have traditionally been applied.

It may occur to the reader that this clarification makes no reference to general systems theory. In fact, the reverse is closer to the truth. The kind of thinking that has given rise to the models used in systems analysis and general systems theory are the natural and synthetic result of decades of careful scientific differentiation, the integration of the resulting body of complex information, and an intuitive or imaginative leap to a new interpretive model. This presentation is based on the assumption that general systems theory is a beginning toward a clearer elaboration of that new model. In discussing social systems, we shall see that the emergence of this model has some interesting relations to the evolution of our social system.

No systems or entities can be clearly defined in such a way as to correspond to terms like "mind" and "body" if these are used in relation to a traditional interpretive context or model. At every stage of its development, general systems theory has shown that there is never a sharp line between any two levels of systems and that the transformation is brought about through changes in aspects of systems or through the modification of relations until the accumulation of these changes results in a noticeable or describable difference in the system as a whole. How much change constitutes a different system is a matter of convention or convenience. Even the most accurate terms are, to some extent, arbitrary limitations or divisions within the unbroken continuum of existence. In our experience, there is a useful distinction between moving and thinking. Because we are able to imagine a variety of responses to a stress, we have to learn that there is a difference between thinking a solution and acting it out. When this simple distinction is considered in relation to the riches of mythopoeic thought, which has played so important a role in human cultures, it is easy to see how the mind–body duality arose. It is often referred to as Cartesian dualism because, in modern times, Descartes made a sharp distinction between extended things and thinking things, that is, bodies and minds. However, Descartes would have had little impact on Western culture if his carefully reasoned position had not fit suitably into the interpretive model that was part of the nucleus of that culture. One might better refer to Platonic or Christian dualism, or both. Ancient Judaism made no such clearcut distinction, but contact with Greek philosophy guaranteed that the distinction would become central in Western civilization.

This distinction, and the others implied by it, leave us with a model that simply does not fit the nature of existence as we are now interpreting it. For example, is electromagnetic energy (e.g., sunlight) physical, mental, spiritual—or what? Is the magnetic field around a magnet physical, mental, or spiritual? Is the flow of electrons through the neurons of the brain and nervous system physical, mental, or spiritual? A slight electrical current introduced into selected parts of the brain will cause one to hear music, remember past experiences, even hear voices. Has the mind or soul then been possessed or controlled by a physical, mental, or spiritual force? Concepts such as physical, mental, and spiritual make it impossible to cope with contemporary information concerning the nature of the human psychic system. If we set aside these prejudicial concepts and direct our attention to the organization, processes, and development of psychic systems, the unnecessary confusion is immediately reduced, and we can focus on the definition or clarification of actual problems. We have been preoccupied with consciousness, but this is a relatively limited aspect of the human psychic system, and a small portion of psychic systems in general. One might argue that self-conscious awareness is the most important aspect of human existence and the one with which we are all ultimately most concerned. Subjectively, this may be the case but, to the extent that it is, human problems and human suffering are increased.

Subjectivism and Alienation

Greek, Christian, and Cartesian dualism culminated in the writings of Jean Paul Sartre. In keeping with good philosophic tradition, Sartre analyzed our conscious life as being in search of the authentic self. He followed a well-worn path in describing the roles that an individual plays in certain social situations. His heroes become aware of the unreality of man-made roles and long for something real and authentic. They are seized with nausea and anxiety when they realize that they really know nothing of the external world. All that they thought they knew was merely their own meaning imposed on the "thing-in-itself," of which they know nothing. In *Being and Nothingness*, Sartre makes it clear that consciousness can neither be defined nor explained. It is "no thing" or nothing. It is not any of the roles or identities it assumes, and the objects and situations of external experience are unknown as they are or may be in themselves. Thus, when consciousness comes to the hard truth, it realizes that it is utterly alone, that it assumes roles in order to relate to a world of its own creation, that it has no inner describable nature, and that it can never know the real nature of the world in which it participates. The result is first nausea and anxiety, then freedom to accept life as our own creation, and, ultimately, freedom to choose the kind of life we should like to create.

This profound insight also presents the kind of one-sided picture we always receive from "rationalists" who make no use of scientific information and thus are victimized by the logical implications of their cultural models. Sartre follows the age-old philosophic tradition of supposing he can reason or analyze his way to a superior perspective from which to analyze the human predicament. However, according to Sartre's own analysis, his philosohical position is merely his own way of perceiving himself and the world. He has made it very clear that he chooses to see himself alone and free to fashion his own life. Clearly, he wants a world where no responsibilities are put upon him except those he chooses for himself. No goals, no absolute moral laws, no limiting ultimate truths. How does one show that such a philosophy is more "true" or a more accurate description of the human predicament than the traditional Christian or Jewish view? If one has no proof, no evidence, then it is a matter of taste dictated by one's cultural inheritance and personal conditioning. In that case, we should all choose the garden of Allah.

Eastern Analysis

Many centuries ago, Hindu philosophers analyzed the nature of man in a manner somewhat similar to that of Sartre but arrived at a very different conclusion. The East has long regarded linear, logical thinking and empirical knowledge as human, transitory, and deceptive. They sought and perfected methods for finding the true inner self or atman. They devised names and categories for various roles, attitudes, and social functions of the empirical person. They included, in their description of this empirical person, their analysis of the whole organism and, to this extent, saw it as an inclusive system. Their search for the true self led them to look behind the system (turn the eyes inward) for that which was permanent and lasting. They found it in the innermost center of their being, but it was a "nothing," and in arriving there all differentiations between subject and object or object and object disappeared into a vague awareness of total unity. The closest approximation, in ordinary experience, are the moments waking from sleep or from the effects of a general anesthetic, when one is aware of being aware but is not really aware of "anything." This state of awareness that is empty of all content is considered awareness of the true inner reality, that is, the self or atman, and is the last stage of consciousness before going into a nonconscious or beyond-conscious unity with this inner reality, which is, of course, also the ultimate reality, Self, or Atman. The proper result of the full realization that one is Atman and that one's empirical self is temporary and relative unreality is described, in the Bhagavad Gita. First, one should attain a detached attitude toward the empirical self because it is not really oneself. It is only a temporary appearance: the whole of one's empirical existence is a role that Brahman–Atman has chosen to play in the cosmic drama. Second, one must do each thing as a gift to God (Krishna, Vishnu, Brahman–Atman) and nothing to satisfy the greedy, glory, selfish ambition of the empirical self.

Hence, since this analysis arose within the context of a philosophical–social system that included a strict caste system, one must do one's duty, to live by and enforce the god-given caste regulations.

It is tempting to move now to Christianity and Taoism and to point out the way in which each has handled the analysis of man's nature and the inevitable realization of the changing, tentative, inconsistent, and temporal quality of existence. However, examples as contrasting as ancient Hinduism and modern Existentialism are probably sufficient to illustrate the general agreement over certain aspects of human existence among those who press the search for individual identity to its limits. In ancient or modern times, East or West, there is general agreement that careful analysis reveals the human empirical self to be a system of roles, responses, or behavior patterns. The prevailing opinion over the centuries and around the world has been that man must have a more permanent and essential nature than the analyzable, empirical self. In the East, it was usually the ultimate universal reality that is the unity and inner reality of all existence. The nausea in modern existentialism arises from the insistence that there is nothing beyond the empirical self except the "nothingness" of the ultimately alone and isolated individual conscious awareness. In Sartre's thought, I alone create myself by choosing what I have become. I alone bear responsibility for my acts and the kind of person I have chosen to be.

From the perspective of general systems theory, these views of ourselves and our nature can be best understood as attempts to resolve the polar tension between our need for relative autonomy (individuality) and our need for system dominance (participation). The one extreme suggests that participation to the point of total identification with the ultimate is the fundamental reality and that the differentiations, specializations, and relations that constitute the existential system of systems is temporary and relative unreality. Hence, the relative autonomy of subsystems must be seen as a temporary condition to be transcended and overcome. The other extreme sees the relative autonomy of the individual as the ultimate reality and any attempt to lose individuality, responsibility, and self-determination through participation, conformity, or role playing as "bad faith."

Equilibration

There is an ancient wisdom that recognizes that everything is a balance between extremes and that the extremes bring destruction. From ancient China or India to ancient Greece, moderation was considered the proper position between destructive and distorting polar extremes. General systems theory expresses this ancient wisdom and the correlative findings of modern science in the principle that every existing thing is an equilibrating system constituted of one or more constellations of opposing forces or processes. Consequently, general systems theory begins the consideration of the human psychic system with the assumption that it is necessarily a subsystem within the whole organismic system, that the organismic system, including the psychic system, is necessarily subsystems in overlapping psychosocial and ecological systems, that the psychic system itself is constituted by subsystems that are clearly definable and analyzable (as is true of the whole organism), that there is no mystery in the psychic system beyond the emergent novel characteristics arising with each new synthetic entity, and that these—and all—novel characteristics are relations of some kind to other systems, that is, bonds or constraints interrelating a new level of system.

Our Changing Cultural Model

There are historical records of feelings of alienation, dislocation, and meaninglessness in every age of cultural transformation. This is in part due to the process of disenchantment with cultural models and myths. The facts and incidents of daily life need a context of significance if they are to remain something more than boring routine. Humans have their individual compulsions and fantasies, but most need some social validation or reinforcement if these are to stand. When cultural myths collapse, individuals begin to question the significance or worth of what they are doing or being. This is consistent with systems analysis. There is a polar tension between the processes by which social systems condition or coerce individual or small group subsystems to conform to cultural norms and the processes by which subsystems eventually force adaptations and/or transformations in the culture or organization of the social system.

In our time, the anthropomorphic cultural myth is suffering one more major transformation. The idea of the universe as a system of systems has been gradually developing for 400 or 500 years. It took over the model of the physical universe with relatively little stir because we in the West had separated ourselves from the physical universe, even the "physical aspects" of our own being. There was a considerable clash over the astronomical aspects of this model, but this was soon rationalized as the system of cultural meanings found ways to accommodate the new position of the earth relative to the sun. However, when this approach began to invade the analysis of living things through evolutionary theory and chemical reductionism, then there arose great resistance. Finally, the same analysis has been applied to the human psychic system and social systems. In every case, the analysis sought to explain the organization and processes of an entity by carefully describing the elements, forces, relations, and movements within the entity. It is not surprising that a theory of systems in general should arise out of the gradual perception of the many isomorphic relations among the various kinds of systems. Nor is it surprising that a general cultural uneasiness, sense of alienation, or dislocation should arise wherever the impact of this new model began to be felt. With reference to humans, the model is neutral. Man is one system among many. We may be of great significance to ourselves and to our fellow humans, but the anthropomorphic link to the universe does not fit. Our sense of participation must now be in the other direction. That is, we stand at the apex, the growing edge, and bear responsibility for ourselves in relation to the systems of which we are made and in which we participate.

Sartre's mistake was a culturally determined emphasis on consciousness as the essential human element. Consciousness is, of course, a novel characteristic of the human psychic system, but to analyze consciousness or self-consciousness in itself is to analyze a function while ignoring the system within which it is a function. This is the same as analyzing the function of a knife without reference to anything to cut or spread. Consciousness is the function of a neurological subsystem of the psychic system, which is in turn a subsystem of the organism, which is a subsystem in a variety of overlapping psychosocial and ecological systems. In terms of this model, we play a very important role in determining the developmental patterns of the future, but we do not have the significance we thought we had under the old anthropomorphic models. This general systems model is just emerging and will, no doubt, encounter great resistance in some places and be ignored in many others, but it will eventually transform the basic cultural myth and our image of ourselves and our society.

Human history provides us with a record of our interpretive models and their transformations. Transformations in the West have been relatively rapid, beginning with mythopoeic models, such as those of the Mesopotamians, Mycenians, Minoans, Dorians, and Ionians. We shall leave the development of changes in the "Western mind" for later, and for the purposes of this chapter it need only be pointed out that, beginning with Greek philosophy, Western thought emerged from imaginative, storylike, anthropomorphic interpretations and world models by the gradual perfection and application of first rational and then empirical methods. As people learned how to apply the laws of logic to their use of symbols, both numerical and nominal, logic became the method of testing the consistency of the meaning of the symbols and their relations through the reasoning process. Grammar and syntax describe the system, the functioning of which logic is the test. Most of the systems we have considered thus far have had clearly definable subsystems, such as atoms, molecules, cells, tissues, and organs. The system of language has the problem of subsystems, such as words, phrases, and sentences, that may have several different definitions, as well as a number of known or unknown emotional evaluations or significations. The possibility arises of several different combinations of meaning from the same combination of symbols. Put differently, one may have a single system of symbols with a set of grammatical and syntactical relations from which several systems of meanings can be derived. If the words (symbols, subsystems) had only one definition, then grammar and logic would be the same discipline. They are not, because one can arrange together words into proper sentences (a function of syntax) and still convey a confused meaning, a variety of meanings, or no clear meaning at all. (Some years ago, the Greek word for signification was anglicized into the term "semantics" to refer to the study or science of the meaning of words.)

However, the West learned that even the most rigorous application of logic to reasoning processes would not guarantee the existential truth of a set of propositions. The philosophical systems of Descartes, Leibniz, and Spinoza are memorials to all those who believed that we could obtain the secrets of the universe and solve our own behavioral problems by manipulating symbols in our heads or on paper. Rationalism, in this light, appears as a sophisticated kind of magic. In mythopoeic times, humans engaged in imitation magic through the use of sacred words. Rationalist thought assumed the human mind to be the true measure of the universe in that it assumed, and often stated, that what makes sense must be true, or conversely, what the mind finds inconceivable cannot be true. The fallacy in rationalism is the failure to distinguish between the symbol and what is symbolized.

The empirical method gradually emerged as a careful process for testing words, definitions, and descriptions against the behavior of the reality they are supposed to symbolize or represent. The correction of our information has not always been easy, even when the method has functioned well. Words and symbols always have meaning within a system of context or meanings, and we cannot change subsystems without some resistance from the system. When the change in a subsystem involves a serious transformation of the whole system or large portions of it, then the resistance can become intense. Further, when the methods of testing become refined and reveal the nature of things to be different from the evidence of our senses, then it becomes even more difficult to overcome the resistance of the system. However, since social systems may be very complex and have many subsystems, there are likely to be many different variations and elaborations on the basic system of meaning. This allows for differential evolution of some areas, that is, systems of meaning, and sets the stage for eventual stress when and if these meaning systems come together.

In the West, it has often happened that some group of investigators or experimenters have developed a new body of information concerning some limited aspect of the universe. While this growth of information remained within a particular social subsystem, it could grow with limited reference to the larger or more inclusive concepts of the culture, and thus with limited conflict or stress. However, because the autonomy of any subsystem is directly limited by its necessary interactions within the inclusive system, there must come a time when the implications of the temporarily isolated body of information comes into conflict with the existing cultural model. The resulting conflict travels through different levels and subsystems of the social system, making slow progress against varying degrees of resistance and demonstrating what the sociologists call culture lag.

For example, in the first decade of this century, Bertrand Russell and Alfred North Whitehead wrote the *Principia Mathematica*. Some 40 years later, more than half the average human lifetime, the so-called "modern" mathematics appeared in the secondary schools and was hailed as the "new" mathematics, and so it was. However, it was also no more than a popularized and simplified presentation of information that was approximately 40 years old. This example illustrates how a body of information can begin with a very specialized social group and then proceed out through the social system. Before it could go very far, it had to wait until each new group had its chance to resist, consider, and then adopt the new model and finally pass the information on to the next group, which would, in its turn, be professionally threatened until the new model was mastered and then join the new elite.

Taking a very long view, a neurological system emerged among the primates that had the capacity for developing an awareness of its own processes. That is, it was aware of its own content of images and symbols, both waking and sleeping. The creative process is one of putting elements of experience together into new wholes or systems. This requires the capacity to be aware of elements of experience and to imagine them, or synthesize their images or symbols, in a new set of relations that constitutes a novel system. This may result in a painting on a cave wall, the invention of new tools, or the discovery of a new use for fire. One of the most fascinating characteristics of this new capacity is that its development is not genetically determined. As Piaget points out in his account of the development of children, and as appears clearly in the historical development of the human psyche, intellectual development is dependent on social interaction. Because the human model begins with other humans and only later becomes the socialized model subjectively modified, it is not surprising that we began with a universe interpreted in the image of human psychic experience. Because this model is the product of social interaction and individual experience, it produces the psychosocial norm or model providing the concepts, interpretations, and evaluations for each new generation. There is generally a heavy emotional taboo against any tampering with the established model.

It is fascinating to note the consistency with which ancient myths describe the struggle between the forces of chaos and the forces of order. The tribal interpretive myths constituted a guarantee against destructive chaos, usually with severe penalties for those who allowed inventiveness to become dangerous deviation. Sensitive humans in every generation have known the fear that, if controls are relaxed, everything will fall apart into total chaos as, of course, it often does in some degree for an individual or a social system. On a larger scale, such "chaos" is often the partial disintegration necessary for the transformation of the system or system of meaning. On a smaller, more local scale, it may well mean destruction or meaninglessness for some individuals, some subsystems, some patterns of organization, or, to some degree, the psychosocial model.

Intellectual development in the West follows such a pattern. We are just now beginning to feel the "culturewide" impact of the empirical method that evolved as a clearly defined method at the end of the Middle Ages and during the Renaissance. That is, there have been a series of conflicts as each new body of knowledge emerged, but they have been somewhat localized. Experts in the information prescribed by the established cultural model and experts in the new information emerging from various sciences have had some notable clashes. For the most part, the general public seemed to go on in its normal way, that is, enjoying the shocking news and then ignoring it in the routine of their daily interactions. In each generation, there have been those who reject the new information, those who devise new ways of rationalizing the cultural model and system of meaning with the new information, those who speculate on a new model and develop a select group of followers, and those who express the despair that is always experienced with psychosocial dislocations or transformations. This is not a new process as we shall see when considering social systems, but there is something new in the way it has been happening in the West.

What is new is the emphasis on self-conscious awareness and the clear definition of the method for attaining it. This means the capacity of the human psychic system to become critically aware of some of its own content and to apply to that content clearly defined methods for testing its logical consistency and empirical validity. The combination of logical and empirical tests with the capacity for reflexive awareness of the content of subjective experience makes it possible for the psychic system to have itself, or aspects of itself, as the object of its own examination.

In a sense, of course, this is nothing altogether new. What we commonly call the moral or evaluative aspect of the psychic system always involves images of both approved and disapproved roles or characteristics. What is new is that this is no longer the private province of the poet or mystic; it is now receiving the same treatment already accorded physics and biology. There is still considerable struggle over the proper method of investigation, and a keen awareness that subjective conditions, as well as distortions of perception and information under investigation are necessarily present in observations the psychic system makes of itself. Any set of subjective conditions and distortions alters the perceptions that one psychic system makes of another.

The important difference is the self-conscious emphasis on the method of verification. Whereas rationalists believed that our "clear and distinct" ideas are necessarily true, modern empiricism accepts as only probable information that is verifiable through some carefully defined method. We are in the process of perfecting methods (i.e., intellectual and laboratory tools and processes) for trapping our own unawareness and distortions into awareness. Obviously, this is a difficult and complex undertaking when applied to psychic systems themselves, but we need not be concerned with the present degree of success in attaining an accepted body of knowledge in psychology. If this analysis is correct, success will come in due time. We need only be concerned with the method and the model that is gradually emerging from the application of that method to succeeding levels of existence. What is here called general systems theory is a general perspective that has been gradually emerging as a result of the advance of scientific knowledge and that will eventually provide a new model of man and his world. This model has already emerged in some areas of the cultural system of meaning, but there can be no doubt that it will be a long time, involving many bitter clashes, before it becomes a model for the culture in general. It may in fact constitute a major transformation or renewal of Western civilization.

Our purpose in the present chapter is to apply general systems theory to the present state of psychological information in order to illustrate (1) its usefulness in revealing the common ground among several differing points of view, (2) its usefulness in revealing isomorphic relations between psychology and other sciences, and (3) its heuristic potential for indicating what can reasonably be expected and recognizing the unexpected novel characteristics of a new level of organization.

General systems theory cannot provide new information, nor can it replace the working models of the investigators and specialists. Their models ought to emerge from their findings. General systems theory can provide a means for comparing models and information among all fields of human knowledge, and this may well indicate directions for further progress. Given the present state of psychological science and the infancy of general systems theory, one can do no more than indicate possibilities. No doubt others will rapidly perceive more and chapters such as this will soon become out of date.

Methodology

Applying the perspective presented here to psychology, certain things come immediately to mind. First, as suggested above, the organism must be taken as a whole and analyzed in terms of its levels of subsystems and processes. (Note that the last sentence is necessary only in the context of psychology, where our perspective is so disrupted by preconceived notions and traditional ideas.) However a psychic system may have been described, from the position of general system theory it is a subsystem of the organism and as such will have definable relations to the other subsystems and to the system as a whole. Second, we have different names for the same system or parts of the same system. We may write about the physiological system, the neurological system, or the system of reflexes and quite possibly be referring to the same thing. Finally, the psychic system is primarily a control system that has the universal characteristics of control systems: a sensory subsystem for gathering information, an information storage subsystem, a selective or evaluative subsystem for comparing information with discriminating standards of some sort either stored or in the form of feedback from other parts of the system, an evaluative system (the same or another) for selecting the best kind of response for the system to make, and an effector system to carry out the transactions (information and energy) necessary for the chosen response. There is no way to describe the psychic system in these terms without involving the whole organism. It would take another book to relate all we now know about human organismic functions to each of these characteristics. For our present purposes, it is enough to provide some general statements and illustrations.

There is, of course, no one separate sensory subsystem, but many kinds of information and many ways in which this information is passed along. I have, at the present time, a significant amount of pain through my shoulders, in my lower back. My hands are stiff, my knuckles ache, and I find it difficult to write. My whole body has a stiff and sluggish feeling, and I would rather go and lie down than do anything else. All this could be summed up, as we usually do, by saying that "I" do not feel well today or, more accurately, "I" worked too long with a pick and shovel yesterday and today "I" am stiff and sore. What I am really referring to is a very complicated situation in which the condition in certain cells of my body, certain tissues, certain muscles, certain neurons, and certain nerve fibers is being reported level by level to higher order systems through higher order information channels and a variety of decisions are being made along the way and various responses are being carried out. There are activities at the level of each subsystem (cell, tissue, muscle, organ, and so on) that must be carried out for the maintenance of that subsystem. The success of the local processes depends on the cooperation of the other cells, the tissues, and ultimately the circulatory, respiratory, digestive, and other systems of the whole organism.

The "I" that feels groggy this morning is the sum total of all that information and all those transactions. Yet, there is no way that the conscious "I" can be aware of all this, anymore than the central command can be aware of every soldier during a war. At each level of a system, there is detailed awareness of the action in that subsystem, but the information that is passed on must be selective so that the next level of control receives only relevant information for that level of evaluating and decision making. All the detail from all the subsystems would present a bewildering confusion. It is necessary to generalize the behavior of the parts of the subsystem into a description of the behavior of the subsystem as a whole. For some of us, there appeared to be a collossal arrogance in General MacArthur's "I shall return." There is the same kind of Napoleonic arrogance when an individual says "I" will do something, or "I" feel something and thinks of himself as a conscious entity with the necessary support systems. This use of "I," in either case, is somewhat justified as a symbolic and synthetic representation of the action of the whole system or the whole organism. In each case, the real nature of the whole has been filtered out for convenience of representation. The critical question remains as to the relative importance of the various functions of the subsystems.

In the West, we have come to regard linear, logical, and scientific thinking as most important. There have been rebellious authors and groups but, on the whole, Western civilization has moved steadily in the direction of believing that the human destiny is to control the world. Aristotle set the road map when he defined a human being as a rational animal. His position is not totally different from the suggestion that novel characteristics emerge at each new level of existence, but the tendency to treat such unique characteristics as the "real" nature of the system leads into all sorts of mysteries, such as "life," "mind," or "spirit." These mysteries result from lack of information, the actual nature of the system being either ignored or unknown. The progress of modern knowledge has been primarily through the analysis of subsystems and subsystems of subsystems, with an emphasis on seeking more and more detail of smaller and smaller subsystems. In fact, this process has gone so far that specialists have, in some cases, found themselves unable to discuss their specialties with others in the same general area, leading C. P. Snow to refer to those who knew more and more about less and less as "educated ignoramuses." It has become very difficult to see the forest for the trees. However, there can be no doubt but that the process of detailed empirical analysis is the only way to remove the mysteries, inaccuracies, and confusion that arise from vague generalizations and cloud over the actual nature and operation of the system.

There is a growing movement in the opposite direction. In the past 20 or 30 years, an increasing number of books have attempted to show the connection between two or more separate areas of knowledge, and interdisciplinary education has become a cliché of contemporary education. General systems theory is symptomatic of this growing interest in an overview and, in its very nature, expresses the awareness emerging from several hundred years of intensive analysis that an entity or a system cannot be adequately understood until it is accurately understood in terms of its actual parts and processes and until it is understood as a whole with novel relationships. The novel characteristics are the result of the synthesis of the parts into a system or whole.

These considerations have a direct bearing on the question of methodology in the study of the human psychic system. Introspection is clearly limited to a tiny part of the whole system and has no way of every discovering what is going on in the rest. Physiology, neurology, and anatomy are empirical sciences, and the human organism has had to be examined like any other system. Hence, experimental and/or behavioral psychology has developed through the careful examination of physiological functions and organismic behavior patterns, that is, the observation of the structure and behaviors of the system and its subsystems. However, this is not enough in the case of humans. This is the only situation in which the investigator can go beyond his empirical observations and ask the system what it is like inside. Of course, we are so impressed with the usual distorted or biased views that come from the inside that many have come to regard them as worthless. Certainly, the history of man has been one long series of opinions that were not usually well supported in fact. Yet, somehow, these opinions turned out to be reasonably good working hypotheses that have aided in producing a very successful and dominant species of system. It is of the greatest importance to understand the processes in the brain, especially the system building processes involved in learning and, hence, in neurosis. It is equally important to understand the corresponding inner system as it is understood by the system's own monitoring system. The latter may be limited and unaware relative to the processes of the whole system, but it is a source of information in the decision-making processes and, in some cases, appears to be the dominant influence.

It appears that, if we eliminate traditional models and accept the human organism as a normal system and the psychic system as a normal subsystem, we can put aside distinctions between physiological and mental and proceed to describe the functions and processes of the psychic system from every available perspective and search for revealing correlations among the various observations. It was not until we began to study the religions of the world, not as revelations of absolute truth but as very important revelations of human concerns and of the dynamics of social system, that we began to see isomorphic relations among all religious sytems. Similarly, introspection may not give us an understanding of the mechanisms and processes of the whole psychic system, but it does give us indispensable information about the processes operating in part of it. Introspective reports such as those of poets, philosophers, and psychoanalysts are available in such overabundance that we tend to disregard their importance or are required to reject them in order to get a little breathing room. If information from subjective experience is really as unimportant as some would have us believe, why do we try so hard to talk to dolphins? We now have the fascinating opportunity of looking at the human psychic system from different sides and discovering the interrelations. With enough information and understanding, human "consciousness" (ego?) may learn to accept itself as part of a total organismic system to which the "self," in self-respect, may then properly apply.

There are fascinating correlations among Freud, Piaget, and Skinner, for example, and the only legitimate issue among those who follow them ought to be accuracy of definition and explanation. Freud stands in the tradition of all those rationalists who put pieces of information together into a novel theory or system. Hundreds of such systems have emerged in primitive, ancient, and modern religion and philosophy. When hard evidence is scarce, human intelligence naturally moves to synthesize or organize what is known or supposed into a usable system. The danger in this process is taking the synthesized system of explanations as absolute truth or fact instead of holding it as a tentative, testable hypothesis. There are real differences in method, and there always will be. As novel characteristics emerge, novel methods of scientific investigation and new scientific terminology will be required. There are still those who believe that there is no need for biologists to use any methods or concepts that go beyond those used in the physical sciences. If this were true, then the central thesis of this book and of general systems theory would be necessarily false. Biology requires concepts dealing with development, growth, equilibration, self-maintenance, species specific and individual specific control systems, negentropy, reproduction, and the like that differ from those required in the physical sciences. It should be expected that different methods and concepts would be required in the investigation of the novel characteristics of psychic systems, and so there are. However, a clarification is required so that our "liberty does not become license." The new concepts in biology all have their roots in the concepts of physical science, and the subsystems or processes represented by these concepts are all analyzable into components that are explainable in terms of the physical sciences. In other words, novel concepts and methods should be related in an integrated and supportive hierarchy of terminology and methods corresponding to the levels of the existential systems.

The disagreement over methodology is closely related to this point. Terms such as "self," "id," "complex," "syndrome," "ego," and "anima" are all intended to indicate novel characteristics of a psychic system, but they have, as yet, no correlation (co-relation) with physical and biological systems. The theories or working hypotheses that have advanced these cocepts as parts of elaborate intellectual systems do not indicate how psychic systems emerge as a novel system for the synthesis of existing physical and biological systems. This, of course, is no easy task, and it is not suggested here that Freud, Jung, and others should have waited until neurology and biochemistry were sufficiently developed before they began trying to help the emotionally disturbed. However, one can take from general systems theory the suggestion that the conclusion of these psychological researchers should be regarded as nothing more than working hypotheses until the step-by-step correlation from physical to biological to psychic is clear. In this regard, it would appear that experimental and speculative psychologists need each other and that their apparent incompatability lies in their claims and defensiveness, not in their methods.

Chapter 8

Equilibration, Stress, Development, Inertia, and Defense

The term "stress" has been used to refer to any condition that encourages some system to act or react. The justification for using the term in this broad sense lies in the need for a word to refer to anything that encourages or contributes toward an imbalance in a system that in turn, requires an adjustment in the system. At the most basic level of simple stable state systems, we find that atoms, for instance, exist at near equilibrium until the presence of an outside energy source, for example, an ion, requires some internal or external adjustment to enhance the equilibration of forces. The presence of such an external energy source is a stress in that it requires some action on the part of the atom. The action is always in the direction of equilibration, even when it is disintegration. All disintegration of systems must be seen as a movement toward a more stable condition (entropy). Movement in the opposite direction, of increasing complexity and instability (negentropy), is also in the direction of equilibration and may result from the union of existing systems or from the elaboration of specialized subsystems.

In the case of steady-state (living) systems, we find that equilibration requires a variety of continuous processes because of the need for a constant input of matter–energy, the constant need for internal maintenance (including necessary levels of matter–energy throughout the system), the continuous process of disintegration taking place throughout the system, the continuous process of interaction with the environment, and so on. Thus there is constant stress on and within a living system, such that it will begin disintegrating rapidly if the maintenance processes are not constant.

In general, we find that all systems tend to resist disintegration or change with a strength or inertia that is somehow proportional to the strength of the bonds that brought the system into existence. Under some kinds of stress or imbalance, existing systems may bond together to form a supersystem with better equilibration. The stress, of course, may well come from inner requirements of genetic determination. Whatever the source of stress, however, resistance to change is related to the strength of bonding and maintenance processes in the subsystems of a developing system, as well as in systems under stress and threatened with disintegration. This kind of discussion ought always to bring to mind that, in the final analysis, everything is a subsystem in the developmental system that comprises the whole universe. The distinctions we draw are partly a matter of recognizing degrees of difference and partly a matter of convenience.

Learning and Central Control

In the human organism, as in other animals, genetic determination provides for the development of a system of specialized neurons connecting the central nervous system with sensory and motor branches all over the body. While the function of the central nervous system is to monitor the internal and external relations of the organism, it cannot monitor directly every level of internal relation all the way down to the needs of cells within tissues. Considerable filtering takes places before the level of conscious functions in the brain is reached. There are also limits on the system's capabilities with reference to external relations. One need only to return to the discussion of electromagnetic energy in Chapter 1 and recall the small part of the energy spectrum that is visible light, to become aware of the many external relations to which the human psychic system does not respond. We all know of similar limitations with reference, for example, to sound, odor, and pain. There are other limits even within the normal range, such as those having to do with threshold or "just noticeable difference." With all its limitations, the specialized sensorimotor–central nervous system is primarily concerned with the equilibration of stresses arising both internally and externally, and this requires interpretation, evaluation, decision, and action at every level.

Piaget provides a reasonable working hypothesis concerning the development of sensorimotor schemes at the level of basic coordination. Systems evolve through the learning of neural connections between sensory and motor subsystems through ganglia and neurons. Infants and young children must be protected for a considerable number of years because of the slow process of acquiring sophisticated discrimination in the selection or decision-making process. The young of all the higher animals apparently have to map their worlds, and they require the instruction and example of the adult social system to be successful. No other animals, however, require the length of time devoted to the training and education of humans, particularly if one includes primary, secondary, and college-level education as part of the preparation. In primitive and more simple societies, the period of training is considerably shorter. The more complex the social system becomes, however, the more elaborate, differentiated, and specialized becomes the information it contains and the larger the volume of information required of the young for participation in the system.

Humans have a considerable capacity for differentiation and specialization of information, but there is no genetic determination for its full development. Even now we do not understand its full capacity, because it has taken thousands of years of psychosocial evolution to arrive at our present state of development. It is clear, however, that each new generation of humans must be informed, programmed, or taught to develop their psychic systems to the level of a contemporary civilized psychosocial system. The detailed mapping of human systems through modern scientific methods began late in our history, and has proven to be the most difficult. The greatest difficulty lies in the understanding of the highest levels and more complex areas of the human psychic system.

One must assume, on the basis of present information, that the development of the psychic system actually begins with the acquisition of separate simple behavior systems and their integration into more inclusive systems. These systems are, in turn, suitable for integration into even larger systems, and so on. The serious limitation of the experimentalists is that they have not progressed sufficiently in the analysis of this integrative process and the novel characteristics of the resulting wholes at each new level of complexity. In order to move in this direction, they will probably have to direct their investigation toward the "behavior" of psychic subsystems toward other psychic subsystems and the resulting degrees of wholeness and fragmentation in the evolving system. As long as one investigates only overt behavior of the whole organism, the psychic system remains a largely unexplained black box.

We have a reasonably clear idea of how some behavior patterns develop through the acquisition of specific skills. The organism's control system learns, through trial and error, where certain muscles are and how they relate to the movement of certain parts of the body. One can duplicate this experience anytime by learning to use certain remote muscles, such as moving the little toe independently sideways. I remember very clearly the process of discovering the muscle by manually moving the toes and then concentrating on the muscle I could not feel but knew was there. At first, I could only muster an occasional jerk; in time and with concentrated effort, however, it became a clear and direct movement. The same experiment can be tried with muscles around the nose, ears, and scalp, and similar experience accompanies the struggle to master all kinds of skills. There appears to be no established communication pattern between the function of conscious intention and the muscles involved in the desired movement. Riding a bicycle involves the coordination of several skills into a higher order of integration. The communication connections among the skill subsystems must be developed until they become relatively automatic. The conscious mind cannot handle them directly because there are too many variables. Once the skills are integrated into a higher level control system, the conscious mind can serve its proper function of scanning the external world and relating this learned behavior to goals, possibilities, dangers, and the like. The neurons are, of course, already spread throughout the body connecting sensory nerves and muscles with one another and with the central nervous system and brain. Many of these connections are already in use or are ready for use at birth; others must be discovered and developed, and some are never used.

Complex Learning Environment

Discussions such as this are always carried on as if the human organism were a simple acting and reacting organism in a simple stable environment. Such is not the case. Relating to a bicycle or a typewriter is one kind of learning in which the object with which we are concerned is stable and consistent for the most part. Relating to another human, or to several of them, and to a psychosocial system is a much more complicated and difficult kind of learning. The psychosocial environment to which children are conditioned is pluralistic and variable in some degree; that is, the relatively autonomous individuals in the child's environment provide conflicting feedback in response to the child's attempts to find successful behavior patterns. The plurality of kinds of responses is much greater in a relatively open, advanced civilization, such as the West at the present time, than in a social system such as Margaret Mead's Samoa. To whatever degree, children must learn to interpret themselves and their work in different ways as they learn to relate to different people and/or different social subsystems. These subsystems must necessarily be antithetical to one another in some degree, and that degree, in relation to the forces working for integration, determines the amount of fragmentation, conflict, and stress among the child's psychic subsystems.

What is needed are careful studies of the formation and integration of the relatively independent subsystems within the psychic system that arise to answer a specific stress in a specific situation. We know from several sources that there is automatic generalization from one situation to a type of situation. It is possible, as noted earlier, that all learning involves the further differentiation, specialization, and finally integration of generalized behaviors and generalized perceptions of stimuli. Because we all live with psychic systems and within psychosocial systems that are not perfectly elaborated and wholly integrated, the dynamics of human psychic systems are likely to include considerable conflict or competition among subsystems with greater or lesser degree of integration and fragmentation. It would be misleading to suppose that there is any one set arrangement among behavior subsystems. On the contrary, a more fruitful hypothesis is that various levels of behavior subsystems align themselves in a variety of organized patterns in response to different stimulus situations, that is, to different perceptions of stresses or problems. The problem of integration versus fragmentation apparently works in both directions. As in any system, increasing differentiation and specialization increases the degree of fragmentation within a system and requires more clearly defined integration. The smooth operation of the psychic system requires higher and more inclusive levels of integration among patterns of behavior and patterns of perception in response to stresses.

Motivation for Integration

This raises a central problem with reference to the integration of the psychic system. Does some genetic need or tendency cause the psychic system to integrate its behavioral subsystems into more inclusive levels of hierarchic organization or do these higher levels of integration emerge as a response to stress? At this point, the latter hypothesis seems more likely and fruitful. The history of human development suggests that human psychic systems develop (become more elaborate, differentiated, and specialized) in response to stress and in correlation with the development of psychosocial (cultural) systems. There appears to be little evidence for genetic determination of either psychic or social systems. Both kinds of systems appear to evolve through interaction within larger environmental systems and in response to a need for equilibration (resolution of stress). This assumption makes it easier to account for "stabilized personalities" that are not totally integrated and contain conflicting subsystems. Such fragmented or relatively independent behavioral subsystems may have little to do with one another (compartmentalization) until a particular stimulus situation or stress brings them into play and actualizes the potential conflict among them. When such a conflict of behavior systems arises, it may be resolved through clearer differentiation and a more inclusive integrative behavior system. Such reworking of the subsystems may lead to more "realistic" or effective mapping of situations and appropriate responses or may lead to accommodation through greater distortion or fragmentation (repression, displacement, etc.).

Adequate organization of any system requires sufficient integration and control of subsystems to allow the system to work. In the development of psychic systems, each subsystem would be expected to have some degree of autonomy (tendency to go its own way) and, to that extent to be antithetical to the demands of other subsystems and inclusive systems. As in all systems, one would assume that the autonomous and antithetical nature would have to be modified, controlled, or excluded for there to be a higher level of integration. All this suggests that human psychic systems tend to develop relatively independent subsystems in response to particular stresses or problem situations, and that these become integrated into more complex higher order subsystems in response to stimulus situations requiring cooperative activity among them. The stress is not necessarily external to the organism. It may be an internal need or desire. The point is that the stimulus provides the motivation for the trial-and-error repetitive behavior that eventually bonds a number of behavioral subsystems into a higher level and more complex pattern of behavior or subsystem.

In general, higher order systems form when suitable subsystems remain or are held in proximity long enough and in relation to a stressful situation to allow them to bond together in such a way that equilibration of the stress becomes possible. The perseverance of a subsystem, once formed, is relative to the strength of the stress overcome and the basic tendency of all systems to preserve themselves (inertia). It seems reasonable to assume that psychic systems are formed or developed in the same manner. Given the existence of basic subsystems and the presence of a sufficient stress, the organism will struggle to bring into coordination a group of behavior subsystems and repeat the effort until the group is integrated within an established control pattern. In this case, proximity may be expressed, not in spatial relationships, but in drawing together through communication channels. The idea of reinforcement fits well into our model because the relief of stress is the success of the newly integrated behavior subsystem. The variety of training in "successive approximation" illustrates refinement of differentiation and selectivity, whether in perception or in the mastery of integrative behavior subsystems.

Limits of Integration

This line of thought suggests that the psychic system should evolve from simple behavioral subsystems into more complex hierarchies until the whole psychic system develops into a well-ordered totality. This would be one kind of model of a completely "actualized" and "whole" individual. However, this would be an extreme simplification of the actual situation. We must not forget that a system maintains itself by limiting and controlling the relative autonomy of its subsystem. Systems always have the potential for relations that are antithetical to the welfare or existence of the inclusive system in which they participate. That is, there is usually more than one possible response to a stressful situation. Usually, the psychosocial system in which the individual interacts sets limits on which solutions (responses, behavioral patterns) are acceptable and which are not. Here again, there is not just one psychosocial system. Even in Margaret Mead's Samoa, children found it necessary to change households when the situation at home became too unpleasant. Even in a very homogeneous society, there were still enough individual differences (relative autonomy) among persons and small groups (subsystems) to give children (developing subsystems) conflicting feedback information. In a pluralistic society with easy mobility, the diversity of feedback from time to time, person to person, and group to group is much wider. We are in need of greater understanding of the process by which a person chooses among these possibilities.

In general, the choices are to some extent the selection of validating groups (cliques, gangs, or peers) or other individuals (friendships, crushes, or affairs) who assist in equilibrating anxiety (stress) by supporting the behavioral subsystems with which the individual is most comfortable at a given stage of development. To some extent, the choices would also involve the influence of those who represent "ego ideals" or developmental goals. In all cases, these choices would have to do with rewards and reinforcements in relation to the equilibration of stresses.

It is extremely important in terms of this kind of systems model to see these behavioral subsystems and a hierarchy of stresses as relatively fluid and changing with the development of the individual psychic system through maturation and specific changes in small group relations. This model assumes that simple behavioral subsystems may be involved in a number of overlapping higher order subsystems, just as a word may appear in a number of sentences, a molecule may participate in a number of physiological subsystems, or an individual may participate in overlapping social subsystems. Further, these higher level subsystems may participate in a number of more complex behavioral subsystems and so on up to some actual or imaginary highest level of integration. Thus there are kaleidoscopic, constantly adapting sets of hierarchies of various levels of complexity that are more or less loosely united within the total psychic system, but that primarily come into action in response to specific stress situations. This may be a reasonable description of what we generally call "roles." So far, there is nothing to suggest any necessary consistency or congruence among these behavioral subsystems except as they evolve into a higher order subsystem. Any given subsystem may be involved in several higher order systems and, thus, the variety of possibilities for patterns of development must be rather wide.

Parallel Development in Psychic and Social Subsystems

The current state of information suggests a close relation between the development of the individual human psychic system and the nature of the psychosocial systems and subsystems within which it evolves. In general, one would expect to find a continuum among psychosocial systems from the highly homogeneous—with consistent and congruent small group subsystems, well-conformed individual subsystems, and a clearly defined, effective, and enforced control system—to the highly pluralistic and loosely bonded, with complex, multileveled control systems, overlapping and somewhat inconsistent subsystems, and a high degree of relative autonomy among all levels of subsystems. Corollary to this is the continuum of development one would expect to find in individual psychic systems in relation to various kinds of psychosocial systems. In a simple primitive, highly integrated society, one would expect to find well-ordered individual psychic systems with few critical alternatives. Deviant or innovative behavior patterns would either not arise, owing to the absence of stress, or would be quickly excluded (repressed, suppressed, or forgotten) as antithetical to the individual psychic system or the psychosocial system. If such antithetical elements were to persist, the individual might well be eliminated. At the other extreme, in a pluralistic society with fewer constraints in the control systems and overlapping psychosocial systems with conflicting control systems, one would expect to find more adaptable and fragmented psychic systems, which are more subject to stress and have a weaker sense of participation in a satisfying inclusive system. In such a larger, more complex system, there would have to be more specialization in social subsystems.

Specialization means differentiation and increases the possibilities for fragmentation and higher degrees of relative autonomy. In these circumstances, the individual as subsystem may well have to acquire a variety of very different roles in order to participate in each of several social subsystems. These roles are subsystems of the individual psychic system and are not necessarily integrated into a more inclusive psychic subsystem. In fact, the increased relative autonomy of social subsystems—for example, peer groups, family groups, business groups, or country club groups—may well correlate with relatively autonomous psychic subsystems in the participant individuals.

Basic Needs in Understanding Psychic Organization

The foregoing brings us directly to a consideration of the basic necessities of psychic organization. The most fundamental model of information input (receptor subsystem), decision (selector subsystem), and responsive output (motor subsystem) must be applied to every level of subsystem in the hierarchy of a psychic system. We must assume that as each new pattern of behavior is acquired a subsystem of stored information is formed concerning the nature of the stimuli to which it should respond. Most of these information-storage and selective subsystems have the capacity for learning beyond that involved in their original formation. The stored information and the selection process can become more sophisticated through more exacting differentiation and specialization. These behavioral subsystems must sometimes be integrated into higher level and more complex behavioral subsystems, for example, the combination of skills involved in riding a bicycle. It follows that a new level of information storage and selective subsystem exists, corresponding to this higher order behavioral subsystem.

General systems theory leads us to examine, when methodology makes it possible, the way integrative controls bond together a number of selective-information subsystems into a higher order selective-information subsystem. Highly important in the understanding of human behavior is the analysis and understanding of the degree of congruence among the various levels of control systems. Thre must be some degree of antithesis between one level of control and another, and among "peer" subsystems bonded together into a higher order system. An analysis of the degree of antithesis could be of value in the understanding of some kinds of behavioral problems.

For example, if simple motor skills are never refined (differentiated and specialized) as much as is required by higher order systems, a significant degree of tension or frustration must arise when feedback indicates failure and the upper level control system has no awareness of the cause of failure and no ability to take corrective measures. Lack of development, sophistication, or accuracy at the level of simple subsystem having to do with syntax, grammar, spelling, careful definition of words, and the motor skills used in writing and speaking could produce negative-feedback reporting failure to high level behavior systems and cause the control subsystems at those higher levels to try to "select out" or avoid those behaviors. If, as is usually the case, the psychosocial environment requires these very behavioral responses, an extremely stressful situation is produced that requires extreme solutions, for example, serious limitations on the organism's activities through avoidance or serious distortions of interpretive information in order to make perceptions (feedback) less stressful. We need more information from physiological and behavioral psychology to comprehend the true nature of "defense mechanisms," the neurotic and psychotic distortions of "reality situations."

From a general systems perspective, the outline of the problem is clear, involving at least the following elements.

1. Identification, as far as possible, of the behavioral subsystems at various levels of complexity so that, where retraining is needed, therapy can be directed toward improving available behavioral responses; the accuracy of selection, interpretive, and feedback-information processes; and the integrative communication and control systems binding all of the above together.
2. Identification and description, as far as possible, of the hierarchical development of levels of systems so that the nature of communications networks within the hierarchies can be analyzed.

Specifically, what kind of information is passed from level to level of control systems? At simple levels, it might well be a "yes–no" kind of signal resulting from some sort of "matching" selection process. The conscious mind receives only the barest minimum of information up through the systems of systems, and, in light of the new "split brain" investigations, what Freud identified as the unconscious appears to comprise most of the control subsystems of the whole organism. Finally, how much "awareness" does the communication system provide to higher control systems regarding problems at lower levels?

3. Identification of types of specialized subsystems, that is, subsystems that perform the same or similar specialized functions in different behavioral subsystems throughout an individual's psychic system.

For example, Freud's identification of the superego is clearly important in directing research toward a careful analysis of a high level control system, although it is an early generalization that oversimplifies the problem. There are most probably a number of superegos, which may be related in some degree of organization or may be antithetical to one another. In other words, what functions as ego ideal within one constellation of related subsystems responding to a certain stimulus situation may function as id in a different constellation responding to a different situation. There are many control systems for many subsystems, and any one of these may function as an irrational, or even destructive, demand relative to the organization of another. We see from the split brain experiments that separate judgmental systems appear to exist for right and left hemispheres, which are kept in working relation through the communication system of the *corpus collosum*. When this communication network is cut, however, differences appear. Systems make demands upon subsystems through control subsystems, and the subsystems, in return, make demands on the system. Only in some imagined paradise could there be perfect harmony of all these control systems and their needs. In actuality, there are only degrees of constantly changing equilibration. Clearly, there is much to be learned about the specialized subsystems that evolve, at higher levels, to monitor the equilibrating processes. All control systems are, to some degree, possessed of ideals (models), such as "judgments," "punishments," and "rewards." At the level of cell functions, there is selection regarding what kind of thing is admitted to the cell, and defensive responses that may involve the production of antibodies and the destruction of what is considered "foreign," and therefore rejected. Not all control judgments are wholly genetic. Some of my wife's control systems have "learned" to overproduce histamines, increase fluid retention, and constrict certain muscles and nerves. In other words, she has developed allergies.

4. The analysis of system needs and the autonomous demands that subsystems make upon the system and upon the behavior of the organism as a whole.

Studies working with simple stimuli and simple responses are more manageable in relation to laboratory controls. However, the true nature of the psychic system cannot be comprehended until a method is developed for understanding a group of responses that form into a complex behavior system. Once a system is formed, it tends to take on a life of its own, to make demands consistent with its own needs, and to resist any antithetical element, process, or decision. It is this "system nature" that clinical psychology attempts to work with. At this stage, their terms are necessarily imprecise and tend to be poorly differentiated and overly generalized. For example, the varieties of schizophrenia and paranoia have multiplied until they overlap in a great confusion of vocabulary. Syndrome, however, is not a meaningless concept. If nothing more, it has the heuristic value of suggesting that psychic subsystems form and are capable of self-generated behaviors and conflicts with other subsystems, and that they even attempt to dominate the psychic system as a whole. This can be said about most clinical terms. It would be difficult to give an operational and physiologically correlated explanation of these generalized terms, and yet they do serve to indicate system formations, established interaction patterns, and processes. Jung's *anima* and *animus* are closely related to the ancient Chinese yin and yang, and modern investigators are finding a psychophysical basis for this "bisexual" continuum in human systems through the analysis of the right and left hemispheres of the brain.

Demonization and Development

If I may take a further step back into folk wisdom, the subject of demonic possession suggests possible lines of future investigation. Paul Tillich's definition and explanation of the demonic in *Systematic Theology* lifts it out of the realm of disembodied spirits who wander the wasteland and make noises at night. He suggests that, in the Judeo–Christian tradition, the demonic has always referred to a part (in our terms, subsystem) that tries to dominate the whole, in other words, a subsystem attempting to reorder the whole system around its own autonomous needs or demands. This is the need to "play God," that is, to reorient the world to one's life needs, desires, and styles. In systems terms, this is the need of a subsystem to dominate the system so as to become the "leading part," or central control system. The dynamic tension between subsystem autonomy and system dominance appears to be universal. This tension is obvious in social systems in a revolution or a strike, and also exists, less obviously, in the tendency of psychic subsystems to "demonize" the whole system.

Some innate tendency within the physiological structures of the psychic system may exist that pulls some of is subsystems into some degree of loose organization. However, abundant subjective and clinical experience suggests that fragmentation, conflict, and compartmentalization are common phenomena among the human psychic subsystems. One of the problems in teaching is to bring the students to perceive the interrelated whole of the subject matter rather than a collection of particulars. It is even more difficult to bring some students to see the interrelations of a group of subjects rather than as isolated disciplines. In a pluralistic society, the corresponding pluralistic development of behavioral subsystems (roles) with conflicting evaluation and control systems increases the amount of fragmentation and the opportunity for loose organization, exclusion, conflict, and so on. The central problem is the presence or absence of a "central control" subsystem with a dominant set of standards or values.

This is really the same problem as the presence or absence of an inclusive hierarchy constituting the psychic system. General evidence from religion, folk wisdom, and personal experience suggests that psychic systems exist on a continuum. At one extreme is little organization among loosely bonded subsystems, any one of which may rise to a temporary position of central control (demonization or domination) in response to a certain combination of inner and outer stimulus situations. Such domination, however long, can enhance or destroy the reality orientation (appropriate or effective behavior) of the organism as a whole. At the other extreme is a highly developed, central hierarchical system with an elaborate communication system among various levels of subsystems. The central control subsystem has a reasonably accurate and "realistic" mapping of typical external stimulus situations and of the abilities and appropriateness of behavioral subsystems (role responses). The central control subsystem develops a relaxed assurance that the total feedback and control system can adjust combinations of behavioral subsystems to accommodate variations in the external stimulus situation. This description assumes a developmental continuum, with the first extreme representing the beginnings in childhood and the latter the kind of ideal found in writers such as Fromm or Maslow. We must also assume an intersectional continuum for any actual stage of an individual's development, between fragmented chaos on the one hand and rigidity of perception and response on the other. (See Figure 11.)

As with all actual systems, the majority of functional human psychic systems equilibrate somewhere between these two extremes and somewhere along the previously described developmental continuum. Movement along the developmental continuum is not automatic but results from complex interactions among psychosocial and inner psychic system stresses. Freud's concept of a stabilized personality—that is, one that has established a manageable system for equilibrating stresses—proves particularly useful in this connection. There are, of course, many different possibilities for arrangements of simple subsystems into more complex subsystems (syndromes?). One large subsystem might tend to be dominant, or several large subsystems might be of

Figure 11
Polarities of psychic development.

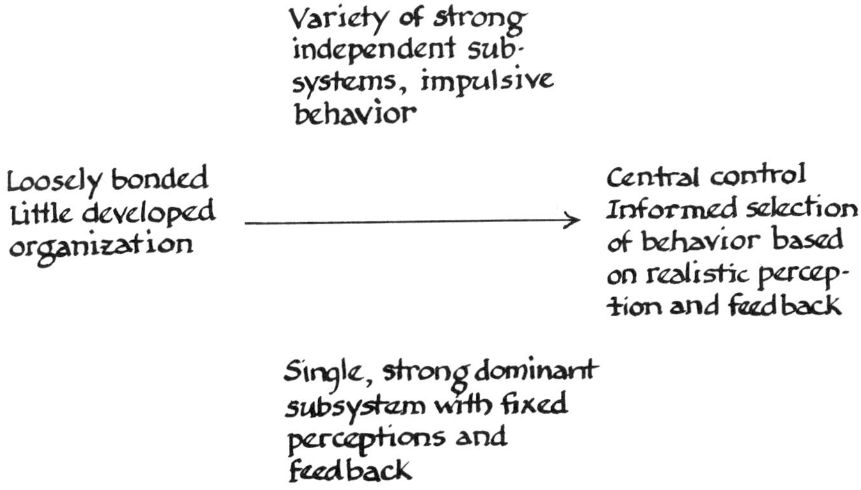

Variety of strong independent subsystems, impulsive behavior

Loosely bonded little developed organization ⟶ Central control Informed selection of behavior based on realistic perception and feedback

Single, strong dominant subsystem with fixed perceptions and feedback about equal dominance. In the former case, "repression" would indicate the presence of subsystems of behavioral responses (maps of past or imagined behavior) that were controlled and judged by the dominant subsystem as antithetical to the well-being of the organism. In the latter case, different subsystems would rise to a position of temporary dominance in response to various stimulus situations and would possibly produce erratic or inconsistent patterns of behavior. Maslow's model of self-actualizing persons suggests a fairly detached awareness of the variety of subsystems and a "realistic" or "rational" choice of the most appropriate among them, an absence of the sudden tyranny of impulsive subsystems responding to stimuli, and a deliberate reconditioning process for the gradual integration of various branches and fragments of the psychic system.

A general picture arises while reviewing psychological theories and attempting to understand them in general systems terms of a gradually emerging system. The application of a systems perspective unites many aspects of different psychological theories, each perceived as concerned with a different level of system or subsystem function. From the perspective of this model, it appears that clinical psychologists use a general terminology that confuses a large number of different subsystems under one term. My guess is that Freud's id and superego will be analyzed into a number of subsystems. In the case of the superego, there may well be isomorphic similarities of a number of control subsystems that function in distinguishable behavioral subsystems. The id will most likely be broken down into a wide variety of demands from resisted, controlled, or excluded subsystems. Filtered information concerning these demands may arrive at a conscious level as restlessness or general dissatisfaction. Displacement, sublimation, and rationalization probably indicate patterns of subsystem development, information storage, evaluation process, and control system elaboration. Thus the levels of investigation represented by the various "schools" of psychology bear some relation to levels in the hierarchies of psychic systems, each needing something from the others to develop a full understanding of the complexities of the human psychic system.

Chapter 9
Psychosocial Interaction, Personal Identity, and Personal Meaning

Ideally, the previous chapter should have ended with a model of a fully developed psychic system. Such a model would have had to be an abstract generalization, such as the model of a cell found in a biology textbook. However, it would have been awkward and misleading to attempt such a description without including the social system within which a psychic system develops. There are single-cell organisms, but there is no psychic system that has not come into being as a subsystem within a social system, that is, a system of psychic systems. Much has already been written about the interaction between the individual psyche and society. What is important in the context of general systems theory is that each is a highly adaptive system and is molded and conditioned by the other. There can be no final stage of development for either because, as long as they exist, they are constantly adapting to changing conditions. Each is, to a large extent, morphogenic (form creating or form developing). None of the psychological investigators have suggested any genetic determination for the development of the content of a psychic system beyond the level of basic biological functions, a few fundamental traits, and simple activity patterns. Nothing like genetic determination exists for social systems except indirectly from the innate needs of the psychic systems (persons) who constitute the social system.

There is, however, one clear distinction between a psychic system and a social system. The psychic system begins with and, in some of its aspects, is structured by the physiological system of the organism. There is a beginning organizational hierarchy from brain to nerve ganglion, to muscle, to sensory nerve endings, and so on, and a beginning hierarchy of areas in the brain for certain functions to take place: for example, the prefrontal lobe is directly involved in abstract thought and vision is located in the occipital lobe. Organizational interaction of specialized areas appears to be interrelated and integrated by special neurological structures such as the *corpus collosum* and various reticula. We need to learn more about the system and subsystem characteristics which are the innate conditions out of which interactions evolve and to which new patterns of relations, new information, and new patterns of behavior will be added. Physiological systems are unique from individual to individual and most necessarily determine some of the range of development for a particular psychic system.

The interrelation of psychic and social systems is the first level at which we can examine both subsystems and systems subjectively and consciously. Here, we need not only the various approaches of psychologists to the individual, but also those of social psychologists, sociologists, and others. Here we can ask the social atom (individual) how it feels to be bound in a social molecule (small group) and gain the full resonance of subjective involvement, because each of us is such a social atom. Philosophy, religion, literature, and drama come alive in a new light when one sees in them some expression of human reaction to or struggle with usual conditions of existence. These conditions become fully available for our examination and understanding at the level of human experience. However, understanding is but a distorted fragment until it is seen as having and developing its very nature and existence within a series of social reticula. Because of these interrelations, a systems perspective requires us to move from an abstract consideration of individual psychic systems through the diametric polarities, bonds, and processes that transform psychic systems into social systems and through which social systems change and are changed by their psychic subsystems.

Diametric Polarities

The relationship between psychic and social systems can be represented schematically if we begin with a few basic concepts and keep the model simple. There are too many variables to fit into a diagram or flowchart. However, we can select a few basic elements and use them to create a dynamic model into which more complicated variety can be introduced later. The basic concepts in systems theory are system, stress, and equilibration. System indicates not only an existing entity, but also those processes that are normal or routine for its established patterns of internal and external relations. Between generation and disintegration, systems consist of many leveled processes maintaining dynamic and fluctuating equilibration between polar extremes. By the term "polar," we do not mean things that are literally opposite in a geometric sense or contradictory in a logical sense, but rather processes that tend away from or resist one another, aspects of systems that are antithetical to one another, patterns of systemic development that interfere with one another, and so on.

The system consisting of the earth and a satellite exists by virtue of the equilibration of the kinetic energy of the satellite and the gravitational force of the earth. A more complex social system, such as a family, exists by virtue of the equilibration of the autonomous needs and developmental tendencies of the subsystems (individuals) and the needs and developmental tendencies of the system (family). The same can be said of the polar tension between system and subsystem at every level of organization. Figure 12 represents the polar tensions between a subsystem and a system. There is tension between the system and subsystem because there is separation and differentiation between the two. Without separate identities, without a sufficient degree of separate existence, there could be no system. The subsystem cannot conform to the extreme of losing its separate identity, nor can it be completely autonomous without destroying either itself and/or the system. If a system exists, it is by equilibrating the tension between degrees of subsystem autonomy and system dominance.

Emerging systemic organization requires adaptation of the subsystem and, under the right conditions, the emerging organization of subsystems necessitates the adaptation of the inclusive system. For example, democracy is supposed to be a social system wherein the emerging needs of the system are held in check so that the system may remain sensitive above all else to the needs of the subsystems. However, because a social system cannot exist

Figure 12
Diametric polarities.

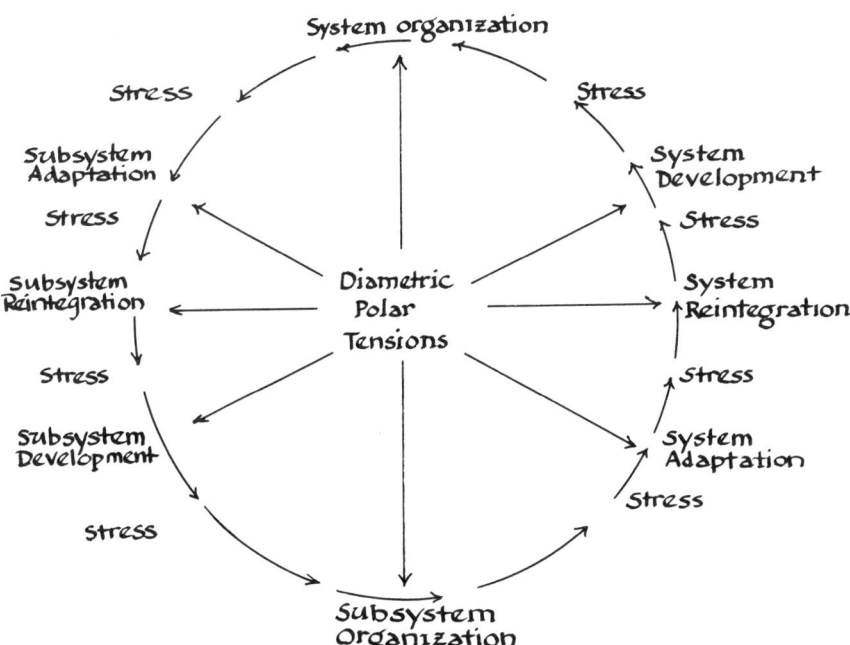

without cooperation and adaptation to the common good, the "commonwealth" must become dominant to some degree. Figure 12 indicates the polarity of the adaptation of subsystems and the adaptation of the inclusive system, between which every actual relation fluctuates. Finally, after a system has undergone some adaptive process, some reintegration is necessary and may well lead to some further system development.

We are a long way from representing the complexities of even a very small psychosocial group. However, the diagram does serve the purpose of emphasizing the dynamic interaction and movement in the development of an individual (subsystem) and a group (system). Neither can develop without the other, and their union depends on the equilibration of cooperative and competitive relations. It would be wrong to interpret the lines of diametric polar tensions as absolute in any sense. They are intended to represent the need for constraints within a control system if the mutual development of subsystem and system are not to lead to disintegration. The main problem with the model is that it is too simple. We have no representation, for example, of the levels of interaction that develop through peer relations. There is nothing in this model to indicate the overlapping and interlocking systems in which any subsystem (individual) participates; the limits of two dimensions make it difficult to illustrate several levels of complexity at the same time. One can put flesh on this abstract wheel and bring it to life by considering the dynamics of the family with growing children. A mother and father pass through stages of adjustment that correlate with the maturation of their children. The family system must constrain each subsystem and require adaptation to the needs of the system if it is not to disintegrate. In fact, a normal degree of disintegration takes place within a family when some of the subsystems (children) develop a degree of autonomy and a need for development that overpower the bonds that held them within the system.

Relations and Identity

In earlier chapters, it was emphasized that systems must be described as existing in degrees. This is particularly true of psychic and social systems, because they come into being by degrees as bonds are formed creating elements of a system that does not yet exist. Learning follows this pattern, as do international relations. The principle at work here is that to be is to be related. This principle also works in the opposite direction: to be related is to be. When individuals form relationships, they begin to exist for those relationships. This may be explained in terms of what the sociologists call roles. If we have different sets of relations for different relationships, we exist differently for each relationship, and that there must be correlative psychic subsystems for each of these relationships. This is no great revelation. Social psychologists and sociologists are providing us with a wealth of information concerning role formation, role development, and the evolution of group formation. Bonds form that establish roles, constraints, and control systems.

Even without social science, we all know that we play different roles in different situations. From a systems point of view, there are two implications worth emphasizing in terms of the meaning of identity for a human psychic system. The first is that individuals find their meaning and identity (psychosocial existence) through the relations (roles) they establish within a variety of social systems. The second is that individuals develop meaning and identity through the integration and organization of their separately developed behavior subsystem.

The adequacy of a psychic system is not simply a matter of inner integration and development. Success is rather a matter of the way in which the kind and level of development in the individual correlates with the social system. A social system may call for roles that are contradictory in their ethical implications, but if each has its proper and clearly defined situation, there may be no problem, and individuals with these contradictory roles may be very well adjusted. This is a common situation in all societies, and in those that have needed rationalizations suitable interpretations have apparently resolved the actual inconsistencies. In every civilized society, at least, there have been those individuals whose inner integration or development has caused them to "see" differently and to speak out against the self-deception involved in these interpretations. These individuals were creative in that they synthesized a different interpretation, an interpretation that may have been advanced, just different, or actually retrograde. In any case, the social system determines whether these people are considered leaders or deviants.

Problem of Self-Motivating Development

The foregoing raises a fascinating problem that is at the heart of the nature of systems. To what extent do systems have a tendency to develop once started, and to what extent do they develop as a result of stress and adaptation? In the present context, the question is to what extent there is an innate tendency for a psychic system to integrate itself and develop higher levels of complexity, and to what extent these are merely adaptive responses to stress. In the context of social systems, it is related to the difference between primitive systems, which change and adapt very slowly, and civilized, urbanized societies, which appear to go through stages of development. The evidence points to the conclusion that both psychic and social systems develop or change in response to stress or crisis, at least at early stages of development. The question remains whether there is a point at which such systems become self-moving, and if development and integration become systemic at some level of complexity. As we shall see, at a certain level of self-conscious awareness, a psychic system becomes a self-developing system. That is, it does not require any external stress to serve as the stimulus for adaptation, integration, and development, and experiences stress in connection with every perception of an unactualized possibility.

This stage of development is directly related to "objectivity" and the presence of variable mappings (interpretations and perspectives) of situations. At this stage, the psychic system becomes aware of the relativity of every interpretation (mapping) and the need to identify those variables that provide the supporting conditions for any given map to give a useful interpretation. This level of operation can be abstractly described in the ideal of the scientific method. Because a psychic system cannot be completely detached or escape the biases of its own interpretive schemes, its highest level is an awareness of those biases as one set among many. A psychic system can acquire mappings (perspectives) from other psychic systems and from examining the mappings of different social systems. Of course, information and perspectives (mappings) are never taken in by a psychic system without being modified to some extent. Some kind of adaptation is necessary whenever information is taken into a system. The adaptation may take place in the receiving system, in the information, or, more likely, in both to some extent. When information is assimilated to an existing pattern (structure for information storage or mapping), the information will probably be modified to fit existing categories. In time, it may well be further modified (rationalized, suppressed, distorted, or repressed) so as to leave the existing psychic structure relatively undisturbed.

There is a degree of objective detachment that, while neither complete nor absolute, constitutes distinctly different perspectives and behavior patterns from those that are situation bound and set in a single interpretive context. This detachment appears to be directly related to the capacity of the human brain to rehearse abstracted versions of behavior patterns in a variety of combinations without actually carrying out the acts represented. Beyond this level, the information system in the brain develops symbols. As in mathematics, these symbols reach higher levels of abstraction until the only behavior pattern is the manipulation of the symbols. For example, the Greeks could arrive at the idea of the second and third powers through squares and cubes; beyond that, however, the powers are a function of the relations of abstract symbols. Most of higher mathematics concerns functions of relations among abstract symbols.

We can best describe this new perspective as a new behavior pattern or model of operation within the information system of the human psychic system, a pattern of seeking out alternative patterns. Pure science, pure mathematics, or game theory are clear examples of human psychic systems seeking alternative patterns. So, too, may be creative activity and invention; that is, the problem to be solved or the stress to be relieved may be generated by the system itself. At this level, the human psychic system is programmed to obtain satisfaction in finding problems to solve and in creating new sets of relations among everything from numbers to sounds. The central stress within the information system is creative manipulation and the discovery of alternatives. Given a suitable social system, psychic systems do become self-developing in some degree, although further elaboration of this point must wait for our consideration of social systems.

Personal Identity

Although much has been written about "personal identity," the term remains vague. It is one of the terms of which those concerned with strict definition are highly suspicious. In recent years, the term "identity" has been used and abused in such combinations as "identity crisis," "true identity," "search for identity," and "loss of identity." Yet it still does not have a clear or widely accepted definition. The subject is not without relevance for contemporary psychic systems; however, because we have no clear working definition of the term, we cannot be sure what information is being transmitted by one psychic system or what information is being received by another. Similar terms are used to describe some modern art; that is, artists are expressing something they cannot or do not choose to define, leaving the viewers to their own interpretations. A similar process is at work in some kinds of poetic and religious writing; in these cases, emotional or aesthetic response is the primary concern. In psychotherapy, one is often struggling with an emotional response that is a complex pattern of behavior and must relieve this stress by learning a more effective pattern of responses. Usually, there is need for more accurate interpretation of incoming information and coordination with a more appropriate response. Clear definition of the problem is one method for beginning a process of change, but it is not the only method and is often a diversionary tactic employed to avoid the real problem.

The question of personal identity may be raised in connection with logical definitions or with feelings of relatedness. In general usage, personal identity refers to more than the activities of the conscious mind. In the model being developed here, consciousness is a novel characteristic of a subsystem of the psychic system. It is an extremely important subsystem with a highly valued function, but it is not the whole system, it does not control the system, and it is not aware of the vast majority of structures and processes that comprise the whole system. The significance of the term "identity" often relates to some part or all of the psychic system and, at times, to some or all parts of the organismic system. Thus, it is not surprising that careful investigation reveals the significance of the term to be related in some ways to the dissociation process described by Piaget, the resolution of conflict described by Freud and others, the valuation processes of the psychic system, and the internalization of models or the learning of behavior patterns from social interaction.

The history of Western civilization reveals a continuing attempt to define and redefine the "essential self," (the true identity in terms of the conscious subsystem) and to represent it as residing within the physical organism. The history of Eastern civilizations reveals a similar search but follows a very different path. The two views are not necessarily contradictory. The West concentrated on the conscious mind, linear thought, logic, mathematics, and eventually, science and technology. The East concentrated on the whole person, total experience, the blindness of the conscious subsystem, and the errors in understanding arising from identification of the human "self" with the conscious psychic subsystem. From the perspective of a general systems model, each has emphasized a different part of the whole. In any case, the meaning of the term "identity" has been very different in different cultures. We must conclude that the meaning of "identity" has a cultural base that is modified in various social subsystems, and ultimately modified within the individual psychic system. Clearly, the word has no one objective referent, unless it is being used to refer to the whole individual or organismic system. Otherwise, it has rich subjective significance with the interrelations of psychic subsystems.

The problem becomes more manageable laid out against the grid of a general systems model. We must first specify the kind and degree of existence with which we are concerned by clearly explaining the set or sets of relations for which or in which the object of our concern exists. This is the case at every level of system relations. Whatever one may mean by the term "personal identity," it indicates sets of relations in which it has that particular meaning. However, many or most of these relations exist in levels of the psychic system which are below consciousness.

Identification

"Identity" is not a serious problem in the majority of our experiences in relating to or in defining most objects that exist for us. In preceding chapters, it has been relatively easy to distinguish patterns of internal relations, extranal relations, boundaries, and transmission (input and output). With the detailed application of systems analysis, it becomes apparent that there is a vague area in the process of transmission when input becomes part of the system and output ceases to be part of the system. In most cases, we can settle the matter with definitions of convenience that become conventions or laws for our common understanding. In all these cases, so long as we can ignore the complications of the observing psychic systems, we are concerned with an objective, public, description of relations. In general, we can locate objects (systems) in space and time, and within a larger system of relations. Each has identifying characteristics, even if just a serial number, dent, or color. Once in a while, we experience the frustrating problem of being unable to distinguish between two nearly identical systems (twins, cars, coats, etc.) but, if we can be sure they exist in separate time and space relations, we can keep the problem within relatively simple bounds. It is when we introduce the complexities of psychic systems that the real difficulties arise.

However, the problem of identity or identification is the same for human beings whether it concerns themselves or something other. At the most basic level, information is received and stored in the psychic system as an interpretation of something. It is a representation in image, symbol, or concept of some system. As an interpretation, it is a synthesis of bits of information from various sources. It may or may not accurately represent some actual system with which there are actual external relations. The synthetic or creative processes of a human psychic system are such that the system represented may exist only within the psychic system itself. Such is the case with all novel ideas or designs. Invention, creative art, and new religious or philosophical perspectives all begin with the development of the internal relations of a human psychic system and then may be re-presented in an external system. When patterns of internal relations evolve in the human psychic system into complex patterns that do not represent any external system, we generally refer to them as fantasy, imagination, invention, or creativity.

The structure of a system depends on classification and generalization. That is, identification, through discrimination and selection, is essential in the organization of any system. The process of identifying an individual or a class involves the interpretation of relations and the comparison of that interpretation with previously recorded interpretations. An interpretation is the organization of bits of information into patterns or systems with whch later information can be compared for purposes of discrimination, evaluation, selection, and reaction. This interpretation is always concerned with relations among systems. From our present state of knowledge in psychology, computer technology, and missile-guidance systems, we know that programmed organized recognition patterns are fundamental to all selective decision-making systems. That is, identification implies an organized or patterned image, a concept, or at least a checklist with which systems can be compared. Human psychic systems are creative; that is, they synthesize and possess the capacity for internal systemic development beyond the mere recording and interpretation of experience. This systemic development may be in the elaboration of logical systems, aesthetic patterns, images, or chronologies of events (stories). History and individual experience offer ample evidence that the human psychic system produces overviews providing interpretive contexts that individuals and societies use to interpret their worlds. There appears to be no essential difference between the organization or integration of a system of bits of information representing and interpreting a simple object and a mythological or philosophical world view, except in the levels of complexity and the degree of abstractness of the interpretive system. The latter is most likely built up through layers of generalization, creative improvisation, and patterns of association to higher levels of complexity and totality.

The problem of identification, then, is more than a one-to-one comparison between an information subsystem and an external system. It is an active grasping and manipulation of information coming into the system in order to make it fit somewhere within a large and complex interpretive system; that is, the human psyche actively tries to assimilate incoming information to existing patterns. All we have said concerning the inertia of a system and system preservation applies to information systems. If the human psychic system cannot assimilate the incoming information to an existing pattern, it cannot satisfactorily identify the source of its experience. In order to preserve itself and relieve anxieties, the creative process will rearrange bits and pieces in its attempt to discover a constellation (subsystem) with a sufficient match. If there is "moderate novelty," the psychic system will develop a new or modified aspect to accommodate the new experience. The accommodation will, of course, be made with as little change in the higher levels of the interpretive system as possible.

From the foregoing, one can suggest that identification is a process of comparison through which a human psychic system finds a comfortable or workable correlation between some source of incoming information and some aspect of its information system. In mapping an individual's world, the psychic system organizes itself into systems that are workable representations of, or symbols for, the systems it interprets to be the source of information input. In other words, the psychic system recognizes the source of incoming information as the interpretive subsystem which it correlates with that information.

Our references to representation or symbolization do not necessarily indicate specifically conscious, logical, intellectual, or rational forms. Psychic systems must be taken as wholes. "Meanings" and "interpretations" exist in nerves, muscles, and behavioral response systems far below the level of conscious awareness, language, or logical relations.

Recognition

Identity is established through identification and recognition (i.e., knowing again). Internally and subjectively, personal identity is the way the psychic system knows (has mapped) itself and thus recognizes or knows itself in its ongoing processes and interactions. The psychic system must have interpretive subsystems (patterns or models) through which it "recognizes" itself. We "feel" our identities are intact when behavior patterns responding to stresses "co(r)respond" with the models (subsystems of information) we use to identify ourselves.

The central problem in the meaning of "personal identity" lies in the assumption of the unity of the psychic system. Although both East and West thought of the spirit, or essential self, as simple, everlasting, and perfect, there is only an abstract notion to support this concept. As long as the term "identity" is attached to the ideas of simple unity and spirituality, it is simply a synonym for "soul," "spirit," "self," and the like. And though in the psychosocial sciences and existential philosophy we have become accustomed to the idea of social roles, little has been written concerning the degree to which what an individual knows about himself is contained in the models of these roles as responses (behavioral subsystems). Identification of our "selves" is directly related to the information we record about responses (roles) as they are evolved. This information is of a variety of kinds, at various levels, and related to different control systems throughout the system. Our sense of identity is related to the ways the various parts of our bodies feel and function. Something is wrong with my "identity" when my leg does not move as it usually does, that is, when its movement does not match stored information. On the other hand, the pain in my left shoulder, although a definite stress, is actually part of my identity. It is "just that darn pain in my shoulder again." It is the "again" that matters, indicating re-cognition, or matchup, between incoming information and a pattern of stored information. The psychic system is built up as it adapts to both internal and external stresses, building up subsystems of response for specific situations, which are then generalized into a class of situations. A particular decision-making subsystem at some level of the psychic system develops a set of criteria (relations) by which that situation or situations of that class are recognized. When the situation is identified, the decision-making subsystem "knows again" (identifies and activates) the proper chain of responses in the learned behavior pattern. The critical question is whether there is an all-inclusive unity that develops and incorporates all subsystems of specific behavior patterns. Informal, subjective, and clinical observation point in the opposite direction. One would expect simple behaviors to be synthesized into large more inclusive behavioral subsystems as needed and developed to relieve stresses. There is little, however, to suggest that such subsystems develop into a single, inclusive, integrated whole. It is common to experience the "coming together" of ideas or elements into a new perspective. On the other hand, it is equally common to be bothered by inconsistent, and even contradictory behavior. The dramatic changes that can result from the effects of drugs have been widely publicized. Many who have learned to be aware of formerly unknown aspects of their psychic system have been impressed with the variety of different subsystems (roles, personalities, or selves). Clinical records support these impressions by presenting a seemingly endless series of accounts of fragmented psychic systems.

Even with conscious intellectual behavior, unity is something to be attained. Any course in philosophy presents ample evidence that learned patterns of thought are often inconsistent and contradictory. The attainment of a reasonably consistent philosophy of life and of the world is a rare and difficult process. When such a view is attained, there is no assurance that the actual behavior of individuals will be consistent with the ideals of the intellectually accepted system.

If the psychic system is fragmented or, at least, not totally integrated, in what sense is it a system? In the absence of verifiable information, one must assume that the unity lies in two aspects; that is, it is a system in some degree but not totally. The subsystems all belong to the same organismic system and function within the same physiological subsystems. Thus, what affects the whole organismic system—or any part of it—may, and probably does, affect every subsystem in some way. One must assume that, even in the case of multiple personalities, the subsystems interact at a level below consciousness. The interaction appears to include mutual rejection and a power struggle. Rejection implies awareness and interaction even if it is unconscious.

The second element of unity or integration is the sharing of perceiving, selecting, and controlling subsystems at various levels of the system. The sensory system is common to all behavioral subsystems, as are the neurological and motor systems. The gradual development of the psychic system is a good example of the value of understanding systems in terms of degrees of existence. There may be other aspects of unity, but his will satisfy our present need.

Loss of Identity

In light of the foregoing, we may conclude that the subjective sense of personal identity does not include the whole psychic system with all of its levels from bodily processes on up and with all of its specialized and fragmented subsystems. We feel our identity in relations or interactions that are familiar, satisfying, and/or acceptable. A feeling of loss of identity would indicate that, for some reason, the usual patterns of interaction are no longer functional. We should expect to have such experiences during some of the transition periods in the process of maturation, or at any time less familiar or consciously unrecognized behavioral subsystems are brought into action by physiological changes or social interactions. When our subsystems are working "as usual," we have no questions concerning our identity. Sartre has written some very clear descriptions of individuals losing their anxiety as they step into a known and accepted role. He calls it "bad faith" and regards it as an "escape from freedom" but makes it clear that, subjectively, these individuals have temporarily found themselves; that is, they know themselves again in a role and a process of social interaction. In all the great religions of the world, "salvation by works" or ritual includes the element of orienting one's sense of identity by the repetition of behaviors designed to restore one's sense of relatedness. Physical activity, or the achievement of limited tasks, often has the same "saving grace": We know again who we are in these acts or achievements.

The psychic subsystem observes the movement, location, and condition of various parts of the organismic system for purposes of control, for example, regulation, maintenance, and response. Most of this takes place below the level of consciousness and functions at limited subsystem levels. Only general information of specific importance is passed on to higher levels. On the other hand, the psychic system observes the organism as a whole, or in larger segments, through usual patterns of grooming and caring for one's physical well-being. A combination of levels of "self-observation" are involved in "watching" oneself in order to improve at some athletic activity. Here, "watching" is more likely to mean feeling than seeing or thinking. In much the same repetition stages that Piaget described in the formation of early simple schemes, the athlete may repeat a motion or combination of motions the "right way" until it "feels right." Higher levels of control are thus programmed to the new behavior pattern.

Self-Consciousness

The greatest interest centers around the human psychic system's observation of itself. Because we know so little about "self-consciousness," we are limited to general suggestions arising from systems analysis. The human organismic system has many layered hierarchies and so, correspondingly, does the human psychic system. If we are not to end in a confusion of nonspecific generalizations, we must relate each term to a level of subsystems or to a function that is isomorphic for all or a certain group of subsystems. It is most likely that terms like "self" and "identity" refer to different aspects of the organismic system in general, and the psychic system in particular, from time to time, person to person, and culture to culture. Ancient literature identified emotions and essential self with various organs of the body, as in the persistent "head and heart" dichotomy. Is true identity the "rational self" or the "feeling self"? This distinction goes back thousands of years with the Chinese yin and yang. It may well be, as noted above, that the distinction is based in the structure of the brain and indicates an intuitive recognition of two high level subsystems in the human psychic system with distinctly different patterns of information organization. The interrelation of these two subsystems is apparently the function of a special subsystem consisting of the *corpus collosum*. Earlier studies located the capacity for abstract thought in the prefrontal lobe, and a revealing relation will most likely be discovered between the capacity for abstract linear thought and abstract image formation. Fascinating questions arise concerning the relation of temporal order, as in a story sequence, to linear order in language and logic. (Dreams seem to show all of these characteristics from time to time.)

There is no one true, final way to locate "identity" or "self" except in the recognition of the whole psychic subsystem and thus, unavoidably, the whole organismic system, as one's "actual self" or "actual identity." There may be some therapeutic resolution of the problem in the direction suggested by Jung, who judged the emergence of the self to take place in early middle age as a transcending perspective with a highly developed awareness of the interrelations and wholeness of the complex human system. There is an interesting correlation between Jung's ideas and some very ancient Hindu definitions of the self (individual) and its relations to the Self (ultimate). Our understanding of "identity" as it is subjectively experienced from person to person may only become clear when related to a personal history and its social interaction. Such a developmental view could explain why an individual tends to identify "acceptable" aspects as self and to repress, ignore, or tolerate "unacceptable" aspects. Such a view correlates well with the theme of self-acceptance that characterizes so many varieties of modern therapy. The poor success record of these therapies may well result from their failure to take seriously the subsystem levels of the psychic system, and to recognize that the rejection of part of "myself" or the negative judgment about my "whole self" may be located in a neurological subsystem far below the reach of consciousness. The disciplines of Eastern Philosophy have recognized for centuries that it takes lengthy and consistent discipline to bring about a consciously intended result in the lower levels of the psychic subsystems. Behavior modification theory may be seen as a modern Western attempt to bring about change at these "below conscious" levels.

A systems theory model suggests that several kinds of therapy must be brought together into a theory that corresponds to the actual complexities of the human psychic system. There is no problem in defining the term "identity" if one operates under the following hypothesis:

> The actual "identity" of a human is the whole organismic system in general and, more specifically, the whole psychic system. Conscious and unconscious experience tend to identify with limited aspects of one or another on the basis of genetic tendencies and patterns of psychosocial interaction and development. The process of maturation is movement toward the integration of the psychic system and awareness of that integration. Therapy should focus on the analysis and location of behavior patterns that block this development and attempt to modify or relieve them by means of the most appropriate therapeutic method available. Therapy in this sense would not be limited to the resolution of specific problems but would be directed toward the fulfillment or development of the possible subsystems, complex hierarchies, and transcending perspective of the human psychic system.

In other words, the variety of definitions of "identity" and "self" and the supposed mystery surrounding these terms are clues for analysis and aids to therapeutic procedures rather than subjects for philosophic debate.

Subsystem Autonomy and System Dominance

We began with simple behavior patterns such as the mastery of the coordination of sensory and motor skills and suggested, following Piaget, that these simple systems are synthesized into larger subsystems in a series of layers, ultimately arriving at the highest levels of control systems in the central nervous system and particularly in the brain. In order to make the above-mentioned hypothesis more understandable, we shall relate it to two forms of a fundamental system tension. The tension is the diametric polarity of subsystem autonomy and system dominance. The psychic system does not come into existence ready-made, nor does it have a genetically programmed plan of development into an integrated system, as do its biological aspects. The elaboration of the system is the morphogenic result of psychosocial interaction. In other words, its survival value is basically adaptability and problem solving. This means that a psychic system will be elaborated into a system that is correlative to the social system and system of problems with which it interacts. It is always tempting to simplify the meaning of "correlative" to something patterned after, but then we would lose the meaning of "interaction." The emerging psychic system will evolve along its own unique line of equilibration between internalizing the pattern of the system and forcing a change in the system through resistance or influence of one kind or another. Both the psychic subsystem and the social system have inertia and are adaptable. The degree of each depends on the relative strength of the bonds of the specific systems involved. As we shall see in a later chapter, social systems are attracted and repelled in a similar pattern of equilibrated tension and may or may not form a supersystem. Most likely, they will form some degree of supersystem, that is, they will have some bonds and informal indirect control systems.

The same analysis can be applied to the internal relations of the human psychic system. That is, the development of a hierarchy of behavior systems does not evolve automatically into one well-ordered supersystem of all control and behavior subsystems. On the contrary, levels of increasing complexity evolve as needed. This, too, is an oversimplification. There are many forces at work in the process of this development. In terms of a general systems model, a behavioral subsystem, once organized, would possess some degree of autonomy and an accompanying degree of inertial resistance to bonding with other subsystems. Interaction with the environment would be continually pressing some of the behavioral subsystems into more complex subsystems in order to conform to the social system or to resolve some other stress. That is, behavioral subsystems are pressed into higher level syntheses in order to solve problems (resolve conflicts) either according to social norms or as emerging inventive combinations. There are many behavioral subsystems and many environmental situations, and there is less than complete integration among the psychic and social subsystems, as well as numerous inconsistencies among both. We need only refer to the significant difference in acceptable behavior among the members of a peer group, a family, or a church group to make this point clear. The difference in behavioral subsystems (roles) that the human psychic system develops in relation to these groups illustrates the separation and antithetical nature of some subsystems. That subsystems are antithetical, however, does not mean they cannot be bonded in the same system. It is a matter of competitive forces. If the bonding force of an integrative system is greater than the antithetical forces among the subsystems, synthesis will be restless or tenuous. The quality of the synthesis depends, of course, on the equilibration forces at work.

Thus in the normal course of events, neither all possible problems nor all possible developments occur. The possibility arises of continuing fragmentation or partial compartmentalization of many separate lines of development. Again, informal evidence supports the idea that various behavioral subsystems (roles) have no necessary correspondence in their evaluation and control systems. Specific behavioral subsystems must specialize to relieve the stress arising from some new problem. Specialization necessarily involves differentiation, and this combination accounts for much of the elaboration or development of a system. As yet, there is little evidence to indicate the presence of an overall monitoring system that automatically realigns the whole system to correlate with the specialization of a subsystem or to maintain some total logic integrating all of the differentiating subsystems. What evidence we do have points rather to the tension created through specialization and development and is directly related to the polarity of subsystem autonomy and system dominance. Psychological defense mechanisms can be viewed as attempts to maintain the apparent unity of the psychic system over antithetical specialized behavioral subsystems. That is, the psychic system interprets actual behavior or behavioral subsystems in such a way as to make them appear consistent with other behavioral and evaluative subsystems. Rationalization is the most common example, and probably the one most often used.

Sense of Identity

A central concern in recent years has been the individual's sense of identity, which, in light of the foregoing discussion, is the question of how the individual "re-cognizes" himself. As individuals develop behavioral patterns in early infancy, they are met with certain responses from the environment. The response and the behavioral pattern are generalized, and a bond or relation is established between a type of psychic subsystem and a type of social subsystem. As psychic subsystems develop into more complex levels, one would expect them to form more complex bonds or relations with more complex social subsystems. If we use Piaget's model of gradual dissociation, we may hypothesize that psychic systems store information concerning a psychosocial subsystem that is a perceived interaction pattern between a psychic subsystem and a social subsystem. This hypothesis suggests that, as individuals mature, psychic subsystems are integrated into ever higher levels of psychic subsystems in response to stress produced by the encounter with (the perception of) ever more complex social systems. The result is the formation of ever more complex psychosocial subsystems (interaction patterns) within which the psychic system recognizes or knows itself. One would expect to find some correlation between the degree of development in a psychic subsystem and the degree of development of a social subsystem, as well as between the degrees of unity and specialization between psychic and social subsystems. This is simply another way of describing the development of role relations, but it has the advantage of emphasizing the interaction and bonding of psychosocial subsystems that tend to become generalized into types and to function as interpretive paradigms for understanding self and environment.

Of course, this process does not take place in a vacuum. The individual always emerges into an existing social system and into an existing natural environment. There is little to be gained from a discussion of some imaginary beginning of social systems, since they already exist at subhuman levels. New social subsystems do, of course, begin but there are always developments or specializations within an existing social system. This forces the emerging psychic system to accommodate itself (conform), in some degree, to existing information systems and established patterns of relations. As the interpretive psychosocial subsystems become more adequate, more interactions will be assimilated, and accommodation (that is, change in psychic subsystems) will be less frequent. However, as indicated in the diagrams at the beginning of this chapter, the autonomy of psychic subsystems and their integration into more inclusive, more complex novel psychic subsystems will always produce some antithetical side effects that will put some degree of stress on the social subsystem to make its own accommodations. The degrees of assimilation and accommodation in either direction will be determined by the relative strengths of the subsystems and the nature of the bonds holding them together. The antithetical elements in the psychic subsystems of a young adult generally develop into more than the nuclear family can accommodate and result in a transformation of the psychosocial subsystem—the young adult leaves home.

A variety of more or less loosely interrelated identities must be recognized for the individual psychic system. This causes no serious problem until individuals are confronted with a social subsystem requiring a radical accommodation with a different kind of social subsystem. In such cases, individuals may not "know who they are." That is, they cannot recognize themselves in some familiar psychosocial interaction subsystem. On the other hand, development and specialization among psychic subsystems may alter perception sufficiently to disturb the established psychosocial interaction systems and cause a similar failure of recognition. A social subsystem could also put stress on a psychic system to respond with a behavioral pattern that is a forbidden psychic subsystem.

In any of these situations, there may be a number of responses. There may be learning, of course, in which the combination of the familiar and the novel, combined with optimal stress, produces accommodation. Such accommodation includes encoding of new information and the acquisition of new interpretive syntheses, and probably involves some degree of new psychosocial subsystem combining behavioral subsystems, information-storage syntheses, and interpretive models. On the other hand, there may be a variety of more dramatic transformations, resulting in a realignment of hierarchical power relations among psychic subsystems, the activation or assertion of neglected or forbidden psychic subsystems, or the domination (demonization) of the psychic system by one subsystem.

Personal Meaning

The sense of personal identity and personal meaning, at its simplest level, appears to be a familiar stimulus in interaction with a familiar response. Individuals "re-cognize," themselves in the interaction. However, there are many responses and systems of stimulus situations. There are many interpretations (models) and systems of interpretations, and there are many responses and systems of responses. As noted earlier, we have no evidence that responsive and interpretive subsystems within an individual form a single integrated psychic system. On the contrary, there is reason to believe that psychic subsystems develop in relation to social subsystems before they integrate into more inclusive psychic subsystems. This suggests that individuals learn identity and acquire personal meaning in psychosocial subsystems by a complex stimulus, interpretation, decisions, response, interaction pattern. If we keep in mind that social systems have many levels of subsystems, and that they will be integrated or not in varying degrees, we have a model of two interacting systems—one social and one psychic—each with a number of subsystems more or less integrated and the subsystems of each forming interacting subsystems with the subsystem of the other.

The model is illustrated in Figure 13, where, if there is any validity to what we have been suggesting, each double-headed arrow marks the occasion for an experience of personal meaning or significance supporting a sense of personal identity. Individuals know who they are in these psychosocial interaction patterns. The situation is far more complex than terms such as role relation would seem to indicate, and it is just this complexity with which existing models cannot deal. The problem with the general systems model is that it makes one aware of so many variables all at once that one may despair of coping with the problem at all. The nature of individuals needs to be understood developmentally as part of psychosocial interaction systems that have resulted in the "programming," "conditioning," "training," or "education" of individuals to interpret generalized cues of a certain nature to indicate

Figure 13
Psychosocial subsystems.

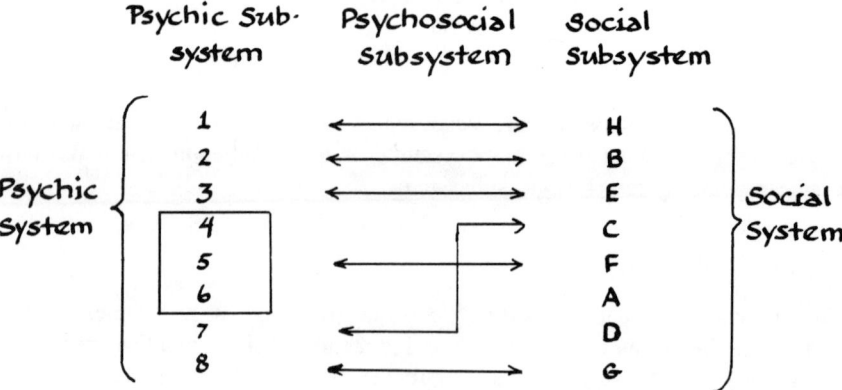

their involvement or likely involvement with or in a social subsystem of a certain kind, and to respond in a certain manner. At the same time, the nature of a social system needs to be understood developmentally as part of psychosocial interaction systems that have resulted in the bonding, structuring, and coding of its internal relations, with certain generalized performance requirements for its subsystems and standards of evaluation for both internal and external relations.

Wht happens, then, is that individuals relate to a social system by means of an established interaction system between a fragment (psychic subsystem) of their psychic system and a fragment (social subsystem) of a social system. Individuals and the social system each have a sense of identity, signficance, or meaning in such an interaction system. This is a feeling of identity in a particular relationship, or a feeling of relatedness through a particular sense of identity. Many writers would agree that we each have many such identities or roles, which change developmentally (transform, integrate to greater complexity, or disintegrate) over periods of time. These identities or psychic subsystems are then more or less integrated into a psychic system.

Transcendent Perspective

We shall not end our discussion of identity and personal meaning with a partially fragmented, partially integrated psychic system adapting and developing in response to internal and external stresses. There is an aspect of the human psychic system that has been recognized and described in some of the most ancient wisdom literature. In the Upanishads of Hinduism, this inner self or atman is the true identity of the individual. It is the transcendent observer of all of the roles (psychosocial subsystems) but is not identified with any of them. Its identification is with the ultimate unknowable ground of all beings, the ultimate Atman. There are interesting similarities between this description and Carl Jung's description of the "self" that emerges around age 40. It transcends and observes the "ego" and other psychic subsystems and involves a sense of reunion in an ultimate dimension.

If we disregard the metaphysical aspects of these two descriptions, there are some interesting similarities with Sartre's description of the "ego." Again, it is "no thing," that is, none of its roles (specific psychosocial subsystems). However, because Sartre has no ultimate dimension, the ego is utterly alone and ultimately alienated. All attempts at providing some sense of union through identification with a particular psychosocial subsystem or with some ultimate realm is "bad faith," or self-deception. Existential freedom comes through the acceptance of the radical aloneness of the ego and its consequent capacity to create the psychosystems it chooses. There can be no doubt that the human psychic system has the capacity to be aware of itself. This involves more than an information-feedback loop combined with an information-storage system; it involves a comparative overview that includes mapping of what are assumed to be whole individuals and total world situations. It is in the perspective of the overview that the identity of the observer appears. Speculating on the basis of a general systems model, we should not expect a single subsystem in which all self-recognition takes place. It is reasonable to expect that the sense of identity as a process of recognizing the familiar begins at the simplest levels, most of which do not rise to consciousness. From cells to tissues, to muscles and organs, to organ systems and systems of muscular coordination, and so on, there are familiar states and familiar processes in which the organism and the psychic system recognizes itself.

In East and West there has been clear recognition of an "observer function" in the human psychic system. This function can develop to the extent of becoming aware of and relatively detached from the various psychosocial subsystems (roles). The critical question in our time is the identity of this "observer function." Sartre set the question clearly, as did the Buddha many centuries before. For the most part, it has been regarded as a substantial and eternal soul. Sartre calls it "ego" and describes it as a "no thing." In most religious contexts, it finds its meaning and identity in relation to an eternal supreme being of one description or another. Sartre leaves it as utter aloneness. The Buddha finds union (meaning and identity) in letting go of the need for specific meaning and identity. "Snuffing out the flame of desire" refers to more than controlling one's passion; it refers to letting go of the need to know. Meaning and identity are significant only in relation to someone who has a need to understand through discrimination and identification. In Nirvana, there is no discrimination, no separation of self and world, or this and that. The true identity of the "observer function" is the realization that the struggle to know, to explain every thing in terms of separation into systems of relations of subsystems, is futile. Life is a cheat. Disintegration is ever at work and eventually wins in each particular system. Hence, the only lasting meaning is no ordinary meaning at all. It is acceptance that the "real meaning" is beyond ordinary meaning. It is a meaning that one can never know with a discriminating mind and that can only be experienced as peace through a sense of "total identity." There are some fascinating implications in the relation of these ideas to the concept of entropy.

On the other hand, Western religions, and some Eastern religions, could not accept conclusions of nothingness or utter aloneness. Through an enigmatic leap, the system is projected into timeless eternity in which the "observer function," variously described, finds its meaning and identity in relation to a supreme being and in a "perfected system." It is not easy to analyze such a system, beyond assuming it to be perfectly equilibrated so that all needs of all subsystems are in perfect balance with all other levels of supersystems and subsystems. Here again, this could well serve as an analytical description of a state of maximal entropy.

Writers such as Fromm, Maslow, and May have apparently set the metaphysics of the "observer function" aside and set about resolving the question of meaning and identity by treating it as a problem of self-acceptance or self-love. In Maslow's description of the self-actualizing person, the "observer function" is very objective, somewhat detached, realistic, and self-accepting. One gets the feeling that Fromm and May are working within a similar context but concentrating more on what hinders individuals from becoming self-actualizing persons, and on finding a cure.

Obviously, we need much more information and a much better synthesis of that which we do know before we can say we have much in the way of reliable knowledge. On the other hand, much can be learned from a more careful examination of the isomorphic elements in the various attempts to define meaning and identity for the human psychic system. Such an examination leads directly into the most basic philospical and moral problems. In the present context, it is enough to say that a sense of meaning and identity for any individual is determined through interaction within a social system. The question of some "genuine" or "authentic" meaning or identity may disappear once the psychic system is clearly described.

The subjective side of personal identity and meaning cannot be discussed without discussing the social system, which is the matrix for individual identities. Chapter 10 will focus on social systems, although the individual psychic system and the question of its identity will remain for us a necessary part of the understanding of social systems.

Chapter 10

Microsocial Systems

That systems exist in degrees is most readily observable in social systems because it is relatively easy to observe their processes of formation, development, and disintegration. When two or more individuals relate to each other, they constitute some degree of system, but the point at which one begins to classify these individuals as constituting a social system is a matter of convenience or convention. Social systems may begin with a few bonds and then may disappear. There is no specific predetermined pattern of development but, when a system does develop, it seems to fall within recognizable limits.

The principle here is similar to Piaget's claim that there is no genetic determination for cognitive development but that, when such development does occur, it passes through the same four stages. There is no natural law driving atoms to add more and more electrons, protons, and neutrons, but if they are added, it is always in a certain sequence. In much of the universe, various kinds of systems develop or do not develop as a result of the interaction between the system and its environment, or other words, as a result of the interaction between the subsystem and its inclusive system.

Aristotelian two-valued logic is of little use here, because social systems are constantly forming by fragments in every casual encounter. Most encounters among individuals do not evolve into fully developed systems, but then, we have no idea what a "fully developed" system would be. Sometimes they develop into a limited friendship, sometimes into a passionate love affair, and sometimes into a family. One cannot say that a social system "either exists or does not exist" or that a social system "cannot both exist and not exist at the same time." They do both and neither constantly.

A sociologist who says that the smallest social unit is the family, is stating a convention, not a "law of nature." It is perfectly proper to set limits for any given study, but it must always be remembered that one is distorting reality by doing so. There is no clear-cut line of demarcation between individual psychic systems and social systems. There are definable boundaries that make it possible to distinguish the one from the other, but their mutual developmental interaction and interdependence is so complete that each is inconceivable without the other.

Social control subsystems are stored in individual psychic subsystems and individual psychic subsystems are informed (programmed) by social control subsystems. The mass of humanity is like the primordial sea, in which macromolecules and elementary life forms were formed and destroyed billions of times before the life process, as we know it, took hold. Social subsystems are continually integrating, equilibrating, and disintegrating. Because of the absence of direct genetic predetermination, social systems may in fact be the most adaptive of all systems.

Buckley, in *Sociology and Modern Systems Theory*,[1] makes the case for social systems being morphogenic (form generating) rather than morphostatic (having a set form toward which a system develops). His presentation is helpful and convincing, but, because he compares social and biological systems, he fails to develop the significance of psychic systems in this picture.

Novel characteristics of new levels of systems make possible the emergence of the next level of development. Clearly, the degree of morphogenesis in social systems is related to the degree of learning capacity in the animals constituting a social system. Primate societies have a higher degree of learning potential than do insect societies. Human social systems are potentially morphogenic to a very high degree because the human individual has a very large learning capacity and very little in the way of instinctive behavior. That is, the human organismic system acquires patterns of internal and external relations as it interacts within its social and other environmental systems. The psychic subsystem maps its own organismic system and environment through such devices as imitation, experimentation, overshoot, and undershoot.[2]

The individual psychic system is highly adaptive. In infancy, it parallels to a degree the cell, which before specialization can be transplanted to any other part of the organism and successfully take on the specialization genetically determined for that area. In the same way, one can transplant an infant from one social system to another and the child will take on the appropriate acculturation, for the individual psychic system will map itself and the new

---

[1]Buckley, Walter, 1967. *Sociology and Modern Systems Theory*. Englewood Cliffs, New Jersey: Prentice Hall.

[2]A map is a representation. It may be a representaton of a land area or, as in mathematics, one set of symbols correlated to (mapped on to) another set of symbols. In order to avoid the ancient debate over the nature of ideas, I shall use the verb "to map" to refer to the process by which systems acquire information from experience. I shall refer to the information subsystems as maps.

social system, and make the appropriate accommodations. Human psychic systems take a long time to develop their potential, and as yet we do not really understand the full potential of this amazingly adaptive, self-conscious, self-motivating system. Many human psychic systems continue active development over an entire lifetime, accommodating new information and new models—all within biological and genetic limits. The psychic system is the link between organismic systems and social systems.

It is the novelty of an emergent organismic systems that appears as a human psychic system, and the novel characteristics of the human psychic system that has made human social systems possible and enables them to be adaptive and form generating (morphogenic). The human psychic system is the connection between social systems and genetic determination. At the same time, it is the memory or information-storage subsystem for the social system.

Bonding

When individuals relate, even casually, they already function within a social system, and have already been programmed with sets of interpretations and behavior patterns. They may, however, continue to relate until their patterns begin to acquire the nature of a bond. *Bonding* occurs when the two individual systems find ways of relating that relieve stress, that is, satisfy needs, desires, and so forth. Of course, there are many more opportunities for such bonding than can possibly be actualized or maintained. There is continuous selection, determined both individually and socially, as to which bonds can be acknowledged, which encouraged, and which maintained as a social subsystem.

The question as to which came first, the individual psychic system or a social system, is meaningless because each has always evolved with the other. It is equally fruitless to try to imagine the beginning of a first social system. If human individuals evolved from a long line of less complex organisms, human social systems must have developed from a long line of less complex social systems. Every individual is born into a social system and every new social system is formed within an inclusive social system, is the transformation of a social system, or is the integration of existing social systems. In other words, every individual and every social system originates in an environment of preexisting psychic and social systems. We step into the process of social system evolution as onto a carousel; there is no beginning, but there are endless beginnings in process.

A General Theory of Human Systems

Once more, we approach an area that has a variety of schools of thought and that covers several compartmentalized fields of study. The problem has been summed up as follows:

A General Theory of Human Systems 151

Civilizationists in the twentieth century have been working out the problems Spengler presented to them. By the 1960's, they had developed a measure of agreement on a model of civilizations, but this did not include the relation between civilizations and their secondary societies or much about the relation between the civilizations and their components, or, of course, about primitive societies.

Anthropologists had, meanwhile, developed theories about the operation of primitive cultures and their relation to civilized societies, but they had not formed a general theory of culture. Sociologists had developed theories about aspects of society including such basic areas as the nature of social relationships, social deviance, social organization. But they had not yet tackled the problem of formulating theories about total social relationships that incorporate macro and micro, historical and contemporary levels.

This statement and the two that follow are taken from the first draft manuscript of Matthew Melko's *Human Patterns*. His goal is to begin the formulation of a "General Theory of Human Systems." He complains that general systems theorists have not come up with productive models:

General systems theorists had formulated ideas about how all systems operate, and from that they have attempted to interpolate some social systems theories; but the fits have been poor, the models have been sterile. They tend to contain elements of a theory, but they aren't worked out.

I shall follow Melko's paradigm for a general theory of human systems very closely in this chapter and leave it to the reader to decide if I have departed from the application of general systems theory. His book is but one more indication of the pervasiveness of the integrative process that continues to spread in our culture. After many years of analysis, specialization, and the gradual disintegration of older cultural models, the normal forces within the system have begun to integrate an overview through high level generalization:

As I encountered first the civilizationists, then the systems theorists, then the anthropologists and finally the sociologists, I had a sense of a mass gravitation toward the central problem. Clearly we have the elements at our command. Several kinds of social scientists with several kinds of knowledge are pushing toward the same goal. It was *obvious* that a General Theory of Human Systems would emerge out of these collective efforts in the last quarter of this century.

I set to work myself to make what contribution I could, to share in the great pursuit of truth. It has been very hard and very exciting. Time and again I have discovered that others have been thinking along the same lines, that Anthony Wallace, Carroll Quigley or Benjamin Nelson had ideas that were running parallel with my own, or that others felt the same as I did about what is lacking in existing theories.

However, this movement is still small compared to the preoccupations of the entire population or even the "intellectual" population. Melko goes on to complain that when he tried to engage his colleagues in this pursuit he was met with indifference and resistance. The emphasis on specialization and analysis has been institutionalized, as it must be for healthy science, but this has led to an "antispeculation" bias. Every enduring, adaptive, steady-state system is constantly equilibrating its internal processes of synthetic integration with those of developmental specialization. Differentiation is necessarily disintegrative to some degree and must be balanced by integrative proceses creating new patterns of bonding. A general theory is an attempt to bond the fragments of specialized knowledge into a new whole, that is, a new interpretive perspective, and represents the other polar extreme from specific research and specialized learning. In Melko's terms, the lack of a general interpretive context is not perceived as a problem by the majority.

However, there is a rapidly growing minority that is producing an ever-increasing amount of literature concerning general theories and integrative perspectives. The tide apparently began to shift about 40 or 50 years ago and has become much more noticeable in the last 10 or 15 years. The fact that I am writing this book on general systems theory and quoting from a book concerned with a general theory of human systems indicates a significant change in our social system: It is equilibrating in relation to a long period of development and differentiation. The old overview has gradually lost its relevance and its ability to integrate the emerging subsystems of information, which consequently appear as antithetical elements, and a new inclusive model or overview has been emerging. This model has grown, and rightly so, from limited generalizations associated with a specific experiment or area of research into wider and more inclusive generalizations concerning a field of investigation. Gradually, the generalizations began to appear similar from field to field, and literature began to be published making correlations between two fields where there was overlap, for example, social psychology and sociology. Today, international societies exist that are concerned with total or inclusive views of man and his world.

We have a cultural identity problem, that is, the old model of philosophical or theological idealism within Aristotelian or Ptolemaic space has been gradually dislocated by analysis and specialization. Subsystems have developed that are largely, if not wholly, antithetical to the *overview mapping* of the inclusive social system. However, that overview mapping exists primarily in the psychic systems of some or most of the individual subsystems, and the change that gradually takes place is evolved in individual psychic systems interacting in small groups, that is, small social subsystems. Those individuals who develop a new area of analysis and correct or enlarge the mapping of some aspect of humans and their world may or may not be significantly changed by their new information. In the preceding chapter, it was noted that the individual psychic system is not necessarily integrated. It may be and often is compartmentalized into various role-related—externally related—subsystems. Similarly, a social system is not totally integrated. Subsystems in colleges, businesses, and professions are pursuing and mapping specialized areas of human information and behavior without any concern for the relation of their specialty to the more general cultural mapping. As Melko points out, this lack of integration is only a problem if it is perceived as such. In other terms, inconsistency becomes a stress only when the two different maps begin to exist for each other through some kind of relations.

This is beginning to happen in various ways. The liberation movements of the late 1960s, for example, brought into play various pressures for a return to law and order, structure, basics, fundamentals, and above all, the old religious model, with its strong pattern of system dominance. At the other extreme, some see the unity underlying the compartmentalized areas of specialization and are trying to fashion a new map at some higher level of generalization encompassing several or all areas of information. These two maps or models will probably come into some degree of conflict at various subsystem levels.

Whether the confict crosses the threshold of awareness for the society as a whole depends on the extent of subsystem involvement and the amount of inconsistency perceived between the older and the newer models. In any case, it is inevitable that, as long as our social system remains viable, it will continue developing, and it necessarily follows that there will be continued analytical specialization on the one hand and holistic integration on the other. One extreme or the other would lead to disintegration.

Melko's book represents part of the integrative process in this social system, and the present book is simply another part. Those of us concerned with an overview are convinced that understanding a larger system as a whole does make a difference in understanding any of its subsystems. Put the other way around, the understanding of any part is distorted, in some degree, until it is understood within the whole in which it participates.

Differentiation

One can begin discussing a system at almost any point of relations or processes because each eventually leads to all others. Melko begins with differentiation and thus accepts the notion presented above that psychic and social systems are always in a preexisting context. "Differentiation" has been a commonly used term in Western philosophy but may not be familiar in current usage. In general systems theory, it indicates the process by which a new subsystem emerges within a system. The process of forming a new supersystem out of existing systems is more conveniently called integration. Although there is no absolute distinction between the two, differentiation suggests further development, specialization, and elaboration within a clearly perceived system, whereas integration suggests the synthesis of systems that had been perceived as autonomous or, at least, separate. Integration will be used here to emphasize the formation of a new system, and differentiation to indicate the formation of new subsystems within an existing system. However, because all integration of new systems is considered differentiation within the universal system, the difference is clearly a matter of emphasis or perspective.

If something is different, it is recognizable as other than something else. Subsystems must be different if they are distinguishable parts, but there are many degrees of difference. Identical cars or twins may be originally distinguishable because they occupy different time and space sequences and relations but, over time, they develop more noticeable differences, they mature and/or age differently because their interaction patterns have been different in some degree.

Differences within a social system are greater, and organizationally role related. Even the simplest social system, the family, contains differentiated and specialized subsystems beyond the obvious biological differences. That is, the family as a social system will have patterns for cooperative effort toward survival. The same is true of primitive societies, which have different functions for different groups contributing toward the preservation of the whole.

Subgroups or subsystems emerge in relation to something that needs doing—in relation to a perceived stress. The stress may be nothing more than the perception of an arrangement an individual would prefer to the existing one. Melko makes this point in relation to the idea of a "leader." Bertalanffy spoke of "leading parts," an important idea in relation to the process of differentiation. A group or subsystem does not have a mind or a psychic subsystem as such. The neurons and cerebral cortex of a social subsystem are located in the individuals who constitute its subsystems. The "mind" or "psychic subsystem" of a social system exists in the "common mapping" of the individual psychic systems, that is, in a common set of symbols, a common context of meaning, a common body of general information, common perceptual interpretations, and a common understanding of organizational patterns and functions.

Incipient differentiation begins every time an individual psychic system perceives the situation in a way that differs somewhat from the norm. Who, then, perceives the "real norm"? Melko's response would be "no-one." The norm is whatever happens that is not perceived by a significant number of subsystems as being different. In other words, there is always variety that goes unnoticed or is tolerated within the constrained variety of the social system. Acceptable behavior is rarely explicitly defined or enforced. Rather, there is a range of tolerated behavior within which occasional variations are inhibited as attention is drawn to them or with the changing moods of the social system. The variety is always present as the seedbed for the development of antithetical characteristics in a subsystem. This type of development creates the "local color" or style in individual psychic systems, families, groups, neighborhoods, sections, nations, or cultures. The tendency toward differentiation is constant, as are the pressures toward conformity (subsystem autonomy versus system dominance), and the state of the competing processes at any given moment is the result of their equilibration or balance.

Melko suggests that an individual who can communicate his or her mapping of a novel goal, process, or model to others so as to bond them into some degree of subsystem may push the process of differentiation beyond the usual limits. This puts a stress on the system, because all differentiation necessarily involves some degree of disintegration for the inclusive system. The stress may be no more than an implicit threat to system control. It may also involve a loss in role relations (services or functions) or include some deprivation for other subsystems. The resolution of such a confrontation is determined by the equilibration of stresses. Management increases the pay of unionized workers because a pay increase is a lesser stress than the loss of profit during a strike. Further, it is stress that can be passed on to other dependent subsystems (the consumer).

If the system controls are strong enough, or the leader is not sufficiently persuasive then the developmental processes will be arrested—the subsystem will have to accommodate the requirements of the system. On the other hand, if the stress of differentiation on the system is less than the perceived alternatives, the developmental process will go on and the subsystem will continue to specialize until it reaches a state of equilibration consistent with the map, model, or goal it began with, that is, until it stabilizes in a "normal" pattern. In this case, the system will have to accommodate the variations in organization, process, or model required by the changing subsystem.

The opportunity for such change varies considerably with the nature of the social system. One of the most fascinating differences is between primitive and civilized societies. The former evolve and change slowly over time in response to the environment, but the common mapping is such that variety is at a minimum and finds little encouragement. The identity of the individual is closely tied to the identity of the social system and the simple organization that defines the nearly identical roles in the relatively few subgroups. The emergence of novel perceptions of stress and novel solutions requires the awareness that alternatives are possible and the perception that they are desirable. The case is quite different with civilized societies, the transition to which is the object of investigation by those who study comparative civilizations.

There is a connection, as one might expect, between pluralism in a society and pluralism in the mapping of individual psychic systems and, consequently, the generation of potential variety in perception and behavior. In a pluralistic society, there are a larger number of subsystems with different mappings and different patterns of internal relations, and an individual psychic system can easily acquire comparative mappings. The very act of comparing requires a transcending perspective and an evaluation process. One must map the mappings. Each of us may possess an individual set of criteria by which we compare and judge among the various maps.

Where do we acquire the superior or transcendent map by which we evaluate the maps that come from the various subsystems in our pluralistic society? There are three possibilities. We use our own version of one of the maps as a basis for judging the others, or we synthesize a new map with new criteria out of the others. A new synthetic map either becomes one more among many or else moves toward a higher level of detachment and objectivity where all concrete mappings are seen as relatively probable working hypotheses. The third possibility is that we may suffer from feelings of alienation and anomie and seek some kind of situation that can put us back inside the mapping of some social system. More accurately, we seek some means of putting the mapping of some social system back inside us as the dominant, meaning-giving pattern for perceptual interpretation, evaluation, and behavior. Religious conversion is often just such an acquisition of a dominant map within a validating community. In 1947, in *Man for Himself*, Erich Fromm predicted a growing mass flight back to conservative (conserving or restoring), nonrational patterns of problem solving as an "escape from freedom." [3] There can be little question that we are living through just such a reaction.

Melko makes the point that humans are small-group animals; that is, they require a small subsystem within which their mappings are validated and their identity recognized or reflected back in some consistent form. Individuals whose "eyes are opened," who stand above the various social maps and see their essential relativity, need an inclusive subsystem as much as anyone else. If it does not exist, or they cannot join it, they begin to seek out and attract others whose psychic systems respond to and identify with their new synthetic map. Their subsequent interaction will evolve a new subsystem map in which each individual psychic system finds a suitable integration with internal relations of roles and institutionalized interaction patterns. Patterns of external relations will evolve at the same time, with accompanying degrees of assimilation and accommodation between the new subsystem and its inclusive social system. With the new subsystem comes further development and specialization of the social system. It may become a fairly stable and institutionalized part of the social system, or it may disintegrate after achieving some goal. However, as such a subsystem continues to form stronger bonds, to answer more needs for participating individuals, and as it increases integration (external relations), it develops larger amounts of inertia. What began as a need to change the pattern of relations and to create something new or different becomes a conserving force for preserving the newly established pattern. Hence, subsystems form internal and external bonds that tend to preserve the newly established pattern.

However, in a complex pluralistic society, there may be many levels of subsystems and considerable breadth, so the integration for the whole system is rather loose. Development and specialization may thus occur that goes

---

[3]Fromm, Erich, 1947. *Man for Himself*. New York: Rinehart.

unnoticed by the system as a whole and by many levels of subsystems. Such development and specialization appears to cause little or no stress until something "brings it to light." The development may have been antithetical to other subsystems or to the system as a whole, but it causes no conscious stress until it is perceived as such. Most complex social systems obviously have many unresolved and ignored inconsistencies among their subsystems. Melko makes the interesting point that when a system's controls are too tight it may increase and strengthen the pressure of antithetical elements in its subsystems. As with all systems, the general tendency is toward some degree of fluctuation in the controls from "too loose" to "too tight" and back again. The responsiveness and stability of the government (control system) will determine the amount of "overshoot" and "undershoot" in its reactions. Too much oscillation (vacilation) eventually brings disintegration or transformation into a more adequate form.

A slightly different situation arises when the inclusive system is confronted with stress that there is no possibility of relieving within the existing abilities of the system. If there are subsystems that perceive themselves as having solutions to the problem, then as the stress moves toward crisis these claims tend to become more acceptable. An individual, a group, or any level of social subsystem can be involved.

The point is that, under "normal" conditions, some potential solutions to problems are judged antithetical to the system and are part of the constrained variety within the system. Under crisis situations, some of these possibilities come to the fore and may be accepted by the system or by enough subsystems. No solution can be adopted without negating many other possibilities and the system transformations each would have necessitated. Hence, every resolution of stress or solution of a problem means new patterns of integration of some levels of subsystems and disintegration of some relations. It may mean the system has to accommodate only slightly, or it may undergo a dramatic transformation. If the resolution does not succeed, the system may repeat the process, or disintegrate a level or two until subsystems become relatively stable systems—for example, when an empire breaks down into nation states.

There is a direct relationship between the shifting of power and emphasis from one social subsystem to another within a social system and the same shifting of power and emphasis from one psychic subsystem (from one interpretive perception pattern) to another within a psychic system. The change in the dynamics of college campuses between 1969–1970 and 1976–1977 would be unbelievable had we not lived through it. Was this a change in a social system, a social subsystem, psychic subsystems, or psychic systems? Clearly, it was "all of the above." Individuals experienced the chaos of breaking bonds and began searching for something to replace the patterns that were lost. Many in the drug generation sought quick insight into the essence of existence but found that existence has no essence except the process of creating, maintaining, and destroying form. Like Shiva, the destroyer who brings life, the most radical rebels of the past decade appear in retrospect as the crest of one of the waves of social adjustment aiding in the equilibration of competing autonomies. The system was too restrictive, so it tolerated the surge of iconoclasts. Now it is too chaotic for many, who are warming again to a new wave of conservers, restorers, and messiahs. There is no way to understand social subsystems, whether individuals or groups, without understanding the tides of the social system. Nor is there any way to understand a social system without understanding the psychic systems that constitute its mind.

Integration

For a system to survive, what has been differentiated must be integrated; that is, relations must be adjusted to the new characteristics of the differentiated subsystems. Control systems must have been adjusted so that differentiation was possible or signaled, and for the system to preserve its integrity, control systems must be adapted to fit the new complexity of the system.

Control systems are relations: Differentiation of subsystems necessarily implies the alteration of both internal and external relations in some degree. Similarly, it can be said that integration refers to the acquisition of more or new relations. We observe this most commonly in the growth and development of our children. In the process, they become more differentiated both in terms of their internal development and in their external relations. Many of the frustrating problems that arise result from the difficulty in developing new relations through various stages.

"Differentiation" is an old term in philosophical thought and has often been associated with the quest for the essence of a thing, or that which makes something uniquely what it is. "Integration," too, is an old term, from a root that refers primarily to a whole, as in "integer," meaning whole number. Integration is thus the process by which something emerges that is recognizable as a separate, differentiated whole.

Integration at the level of internal relations may produce differentiation at the level of external relations. As indicated above, differentiation requires accommodation in external and internal relations until a workable equilibration is established between internal and external integration. "Specialization" refers to differentiation that serves a special function with reference to some system or subsystem. Specializations may relieve stress for a subsystem but cause stress for the inclusive system, peer systems, or subsubsystems. Often, what serves the interest of one system causes stress for others. This is simply to say that what is added to one part of the universe must be subtracted from another, a statement that applies equally well to matter, energy, or adjustment. In fact, the whole universe is a process of processes in continuous equilibration with one another. Without continual equilibration and integration, differentiation and specialization lead to disintegration.

The specifics involved in the integration of social systems have been carefully described by sociologists and are presented in standard introductory texts. Of course, sociology is already the study of social systems in general, and one might expect that, like the other sciences, its findings would be compatible with the theory of all systems in general. When those who study social systems speak of a group, society, culture, or civilization, they are using terms they intend to be applicable to all systems of those classes. Groups tend to differ in some respects, just as cells do, but they must have a recognizable, describably common nature or they cannot accurately or meaningfully be called groups. In fact, it is only vocabulary that prevents the correlation between sociology and general systems theory from being immediately obvious.

Although social systems are constantly forming by degrees in casual encounters, most of them do not persist or develop into groups. Sociologists distinguish between transitory and recurrent patterns of interaction. However, even in the most temporary interaction, some necessary conditions must be assumed. There must be some degree of common mapping and common perceptual interpretations. People from entirely different cultures, with different languages, find the amount of common mapping to be relatively limited. Yet if they experience stresses (problems or goals) that encourage them to pursue some interaction, they will begin to work out some elements of a common mapping. It may be only a kind of sign language indicating objects or needs, but some degree of meaningful communication will take place. They may not communicate directly at all, but behave as some animals do, and begin hostilities. Communication of a sort occurs even in this case, since they begin to interpret each others' behavioral signals and develop a kind of common mapping in the use of weapons, darkness, cover, and so on, for similar purposes. In other words, to be is to be related in some degree. If our two foreigners exist for each other at all, they must relate in a corresponding degree. They may, in fact, develop a highly regulated pattern of warring on each other, and thus develop a significant recurring system.

In another kind of situation, strangers may come together in crowds or audiences and become part of a system that lasts for only a few hours. This is true at the theater, political rallies, baseball games, and the like. These people have a relatively larger amount of common mapping and pattern behavior that enables them to function together in a highly predictable manner. When normal expectations are not satisfied, one may be resentful of the "unruly crowd" or the "unresponsive audience." The common mapping must include common problems, such as stresses, needs, and goals. People come together, even as strangers, to solve problem or relieve some stress. The very existence of a goal is a form of stress. In this sense, a need for relaxation, a desire to "have fun," and boredom are thus all stresses or problems to be solved.

The difference between "transitory" and "recurrent" is not absolute; no such distinctions ever are. A group is usually considered the basic general class of social systems. It is defined as recurrent rather than transitory and as having some kind of social organization. Organization in this case means patterns of behavior governing the interaction of the persons in the group. However, even in the most casual gathering, there are well-established patterns of behavior in each of the persons present. The question is whether there is sufficient common mapping to provide correspondence among the patterns present. We have all been around people who are instantly compatible, as well as those who are instantly incompatible. Every time a young couple goes out on a date, they are exploring in an attempt to discover degrees of common mapping. Those who fall in love may believe they have found a "soul mate" or that they were "made for each other." They perceive what they believe to be a high degree of common mapping, but only time will tell how much of it is in the eye of the beholder.

What causes a group to begin to be identifiable as a social system is the degree to which patterns of behavior, including specific individual roles, become associated with that group and relieve some specific stress for the individuals concerned. As we have seen at every level of existence, there is no hard line between "group" and "not group." Groups come into existence by degrees, exist to some degree, and disintegrate by degrees. Some aspects of transitory crowds are recurrent, and some aspects of the most permanent social system are transitory. It is again a matter of convenience and convention to say that a group is the most simple social system. Even then, there will be many disagreements about whether "a gathering" is a group or not.

Nature has interest neither in the classifications of exact science, nor in Aristotelian two-valued logic. The problem is the same with social systems as with atoms, cells, organisms, or stellar systems. If a pattern of internal relations can be identified, a pattern of external relations, some boundaries, then a system may be described. In the present case, the internal relations would include the patterns of behavior determined by the degree of common mapping and the control systems. External relations would, at least, include the group relating *as a group* to other social entities, or members of the group (subsystems) relating to other social entities or persons as members of the group. The minimal criterion, with regard to boundaries, must be some means of distinguishing members from nonmembers. Even in the absence of all these criteria, it may still be possible to demonstrate that an arrangement possesses some aspects of a system and, therefore, has some degree of existence.

All groups are formed from individuals who have already been mapped within some social system, who already share common patterns of behavior, needs, and goals. Groups are formed as variations of existing social systems for the mutual benefit of the members. Melko suggests that humans are small-group animals and thus need to establish their meaning and identity in small, validating groups. However, these groups will be a modification of the general social system that is more satisfying to the particular persons involved.

Sociologists recognize this when they discuss the conditions necessary for group formation. It is generally accepted that there must be preconditioned components (socialized individuals) with some degree of common mapping (usually some common goals or needs) who interact over a long enough period of time to develop mutually satisfying roles and patterns of behavior. As the system develops, the individuals will become more interdependent (develop stronger bonds) as their need for the satisfactions provided by the group increases or becomes an established part of their psychic systems. At the same time, an increase in resistance to change (inertia) takes place until there develops a tendency toward preserving the group even if it becomes unsatisfactory. This development should be seen on a continuum on which, at one end, strong need and satisfaction pull the individuals together and keep them there, whereas at the other end the group holds even the individuals who do not want to be there. As with all systems, any time the bonds are not stronger than bonds in competing social systems, the group begins to disintegrate.

All social systems contain control systems, which involve selective discrimination concerning internal relations, external relations, and transmission across boundaries. Consciously or unconsciously, the members of a group agree on the way things should be done, such as norms, rules, expectations. The agreements may be formal (written) or informal. Invariably, the rules and expectations include roles (coordinated behavior patterns) for the individuals and an overall ranking system. The roles and the ranking system are the means for implementing the rules and expectations. These aspects may be implicit or explicit, written or unwritten, highly structured or relatively flexible. An actual group is somewhere between these polar extremes and probably a mixture of each. The most formal and rigid social system has its informal and flexible aspects, and vice versa. These possibilities should also be seen on a developmental continuum ranging from simple, informal, and unwritten to complex, formal, written, and structured. These developments are never single layered. In any social system, subsystems are constantly forming and disintegrating as the needs or perception of problems require different kinds of satisfaction, either in accordance with or contrary to the inclusive system.

Problem Perception

Social groups are usually a particularization of the general theses of the social system in response to the specific problems perceived by the members of the group. Making use of Melko's description of a problem as the perception of a more desirable arrangement, we can group all goals, needs, desires, and motivations of any kind together as "perceptions" of the need for a change. To do this, we must allow the term "perception" to refer to all feedback information at every organismic level and to every level of stress. I think this is not Melko's intention. Rather, he seems to use the notion of problem perception to indicate the motivating forces bringing about change in a social system of the sort that leads to accommodation or development. Stress can represent any kind of imbalance that requires equilibration. Routine solutions may be found for many stresses, while others arise for which no satisfactory solutions exist. These require novel solutions, and until such a novel solution is found, the stress becomes a crisis for the system.

However, this distinction is still too simple to do justice to systems. A psychic system need not wait for some stress to arise, but may generate its own stress through inventive imagination. That is, each individual psychic system has many subsystems, each with its own autonomous needs that are more or less controlled as part of the constrained variety of the system.

Psychoanalytic evidence suggests that these stresses are often equilibrated in some indirect way and may provide a pressure in the direction of change. When the psychic system, conscious or not, perceives the possibility of a better arrangement for relieving some of these stresses, the pressure increases. At some degree, the situation is perceived as enough of a crisis for the individual psychic system to precipitate a change in behavior.

Such changes in the behavior of a subsystem may, and often do, cause stresses for peer subsystems, other levels of subsystems, and possibly for the entire inclusive system. If a number of individual psychic systems share a common perception of the problem or stress in need of relief or solution, a group may well form and the change in behavior receive social validation through some degree of common mapping. Sociologists say that groups form when goals or needs are shared long enough to develop some organization of interrelations. Melko's emphasis on problem solving cuts across the customary terms to indicate a degree of stress that motivates social change or development. Often, such change and development involves the redefinition of roles and social ranking for many who perceive themselves as discriminated against by the system.

The critical point in Melko's thesis is that recognition of a stress, pain, or deprivation is not enough to start the engine of the social system moving toward change. It is the perception of a possible solution in relation to a recognition of a stress that produces the movement. Most stresses for which there are no solutions are simply tolerated. If stresses become too great, they lead to the disintegration of the system, sudden or gradual. However, when a possible solution is perceived, action is possible. Such action usually has as its goal a change in role or rank, bringing with it greater social rewards and fewer social disadvantages. This usually involves some degree of further specialization.

Specialization and Organization

The longer a social system continues to function, the more specialization is likely to develop. This may or may not be true of primitive socieities, but it is clearly the case with civilizations. In order for many specialized subsystems to be coordinated, and for their proliferation not to lead to chaos and disintegration, some organization must emerge that will provide the internal relations of a social system. These internal relations will consist of the bonds (relations, roles, etc.) among peer subsystems and a ranking pattern through which roles will be integrated in accord with the emerging pattern of the system, and through which control and feedback systems can operate.

Of course, there is always more than one control system within the system. There are informal controls, informal ranking patterns, and informal feedback systems operating in, through, and around the formally—consciously—acknowledged system. Freud's idea of overdetermination may be applied to every level of psychic and social system. There are many reasons for making any particular choice, and we verbalize the conscious one that best serves our immediate purposes. There are reasons of which we are not aware, as well as some that are completely beyond the realm of consciousness, that is, the many biotic and psychic subsystem levels that influence behavior and decisions below the level of conscious awareness. Similarly, a variety of sources for feedback and control exist in a social system. In a complex social system, one may function as a participating subsystem in a large number of social subsystems, each with its own patterns of information and control.

As with individual psychic systems, there will be varying degrees of coordination and conflict among the goals, interpretations, and expectations of these various subsystems. The subsystems have some degree both of common and antithetical, nonparticipating mapping. Mapping is encoded information and carries with it implications for role definition, role expectation, and interpretation. The variety of mapping provides the opportunity for the production of inventive solutions for recognized stresses. Productivity is also increased by the pluralism of goals, interpretations, and the like, of which individuals are informed by participation in a variety of social subsystems. Conflict between information and control sources is a source of stress with which individuals must come to terms one way or another.

Again we have a complex picture of multiple patterns of organization, including ranking, information, control, and role expectation, which operates from physical, biotic, psychic, and social levels of systems. A careful look at the true complexity of social systems provides a helpful model for the rest of the multileveled universe. A system can exist only if the large amount of variety contained in all the levels of subsystems of subsystems of subsystems can be restrained within an organization that provides sufficient common mapping for cooperation to exceed conflict or competition. When conflict increases in a social system, it may lead to any one of many levels and degrees of accommodation by subsystems, the system, or both. It may lead to the destruction of subsystems and even the transformation or disintegration of the system. One of the more common resolutions is the further specialization of subsystems.

One of the ways sociologists emphasize the importance of organization as a determining characteristic for social systems is by comparing a group and a category. A social category is composed of a number of individuals who share a common characteristic or a number of characteristics. Businessmen, secretaries, and teenagers constitute such categories. These categories are simply generalizations, as are all the words in our language except proper nouns. If the members of a category organize themselves, they become a group or a social system. The process of organization involves the establishment of constraints that become the bonds (internal relations) of the system. The strength or degree of system present in a group can be measured by evaluating the strength and number of constraints that are operative. The group is only one source of constraint for the individual, and all groups combined can not eliminate the novel variety that arises within the autonomous internal process of subsystems. All that can be hoped for is sufficient cooperation to provide for the maintenance of the system. This is best achieved when the subsystems perceive the maintenance of the system as important to their own preservation. The more dependent the subsystem perceives itself to be, the easier it is to have effective constraints and reduce the level of subsystem autonomy.

The perception of the self as highly dependent on the social system provides the strongest base for the exercise of authority in the process of control and organization. The concepts of authority, status, and esteem can only be properly understood as coordination between the mapping of individual psychic systems and a social system. History affords ample evidence that no police force or similar system of control can maintain an efficient and satisfying social system when inner psychosocial controls break down or do not exist. There are always formal and informal sanctions that those in authority (leading parts) can use to maintain coordinated interaction among subsystems.

However, formal sanctions alone have never been enough to provide effective control. The number of social subsystems that have flourished under outright persecution indicates that, when the social subsystem with which individual psychic systems have coordinate mapping is threatened with forced disintegration, the bonds within those subsystems are strengthened. The persecuted social subsystem becomes the source of meaning and identity for the individual psychic systems. More precisely, some psychic subsystems find fulfillment in coordination with the rejected social subsystem and become the dominant aspects of the psychic system. In this condition, other psychic subsystems are temporarily constrained as antithetical to the dominant arrangement; for example, some psychological needs of the individuals find satisfaction and fulfillment through participating in the rejected group. So long as that allegiance dominates the psychic system, the concerns of other psychic subsystems are held in abeyance. The same dynamic operates when a social system is threatened. Social subsystems then set aside other concerns and devote their energies to maintaining the system.

The same thing happens at the level of biotic systems, but with an important difference. At the level of psychic and social systems, the degree of coordination among subsystems and systems is a matter of mapping (learning), accommodation, potential deviation, and constant equilibration. A psychic system is most commonly a loosely organized set of subsystems, often having conflicting needs and goals. It is not difficult, therefore, to understand how an individual can give allegiance to a variety of social subsystems, even those with conflicting needs and goals If one of the social subsystems is threatened, it may well take precedence over all the others.

This arrangement has great advantages for the psychic systems involved and for the social subsystem itself. The more intense the identification a psychic subsystem makes with a particular social subsystem and a limited set of goals, the more the psychic system will overcome its own fragmentary nature. Tensions among rival and conflicting subsystems are suppressed and a kind of unity emerges under the authority of a commitment to a specific system and specific goals. There is a close connection between strong allegiance to a social subsystem and the unity of allegiant psychic systems. Thus, a threat to the social subsystem is a threat to the unity of the participant psychic systems. The disintegration of a social subsystem may well mean exposing involved psychic systems to the dreaded chaos of a fragmented psychic system, with the resultant loss of the sense of meaning and identity. This is not the only way a psychic system can find unity, identity, and meaning, but it is probably the most common.

A social subsystem need not be threatened to be useful to individual participants. Although the system under threat may provide the most dynamic, clear example, every social subsystem acts to some degree as "us against them." That is, every social subsystem has a degree of relative autonomy that necessitates some degree of defense of its own against the system. In our present cultural situation, it is most common for individuals to find a reasonable sense of unity as various psychic subsystems (roles) participating in various social subsystems (groups). There need not be a high degree of coordination among the goals and expectations of role–group arrangements as long as there is no serious conflict. Some people commonly manage a number of disparate involvements.

What is significant is the relation of the mapping of a psychic system and a social subsystem. The more a psychic system maps itself—that is, recognizes itself as a part of a social subsystem—the more authority the social organization has over it. The more intense the sense of unity, identity, and meaning this participation provides, the more sacrosanct the structure, processes, and mapping of the social subsystem become. Organization requires coordination of roles, ranking, and leadership. Those individuals (psychic systems) or groups (social sub-subsystems) with positions of authority in the social hierarchy have status and esteem as special parts of the sacrosanct system.

To reiterate, it is important to realize that the sanctity of the social system is derived from the dynamics of the individual psychic system. The awe, dread, guilt, fear, and shame that deter most of us from certain behaviors originate within our own psychic systems and are projected onto appropriate roles and social systems as we become involved. Nothing is so lacking in numinous qualities as an exposed Wizard of Oz or a god with exposed "clay feet." The numinous experience of the individual psychic system usually searches until it finds another suitable repository for its projection.

As will be seen in Chapter 11, at certain stages in civilizational development it may be difficult for some individuals to find a suitable social subsystem or a suitable object for numinous projection. These are times of alienation, anomie, and despair, because to be without social bonds, or to have only "secular" social bonds, exposes the individual psychic system to its own lack of integration and its own isolated meaninglessness. Meaning and identity exist in relations and nowhere else. Thus when a psychic system finds participation in a sustaining social subsystem, it regards that social subsystem as the authority that maintains order and unity. Whoever participates in the internal structure of the social subsystem tends to have authority, status, and esteem. This is not absolute, of course, but the position tends to be respected. A capable individual can use a position in the social structure to exert more and better control than an entire police force by wisely making use of the coordination of psychic and social mapping.

It is often suggested that social systems must have rewards and punishments to use as sanctions and encouragements in providing a control system. These rewards and punishments are only truly effective if there is sufficient correlation between psychic and social mapping. The record of consistent failure that has followed every attempt to create a utopian society, however, indicates that rewards and punishments are necessary to reinforce proper behavior, even in a well-coordinated society. The reason for this is the degree of autonomy and inventiveness possessed by psychic systems. Because there are many different ways to arrange relations and rewards within a social system, it is to be expected that some individual psychic system will map alternative arrangements much more to their liking than those sanctioned by the social system. Cain very quickly imagined a world in which the favored Abel did not exist; in fact, usurping the position of a favored one is one of the more common solutions discovered by individuals and groups. For social systems to avoid the chaos of "each living in accordance with his own conscience," there have to be constraints on most of the self-serving creative solutions of subsystems.

The most effective rewards and punishments are often informal and personal in nature and directly involved in the mapping of individual psychic systems. Actions that deprive individuals of relations on which they depend, or that adversely affect their self-respect, sense of identity, meaning, or self-esteem often prove to be more effective than more violent but external forms of punishment.

Inventiveness and subsystem autonomy leading to tension between subsystem and system are not the only forces in need of constraint. What sociologists call role expectations refers to the interpretation of an anticipated interaction pattern. However, interpretations are highly individual from subsystem to subsystem, and between subsystems and the inclusive system. These differences lead to confict and competition that must be resolved. A social system that is well integrated and has adequate control systems will have an established pattern for settling such differences. Even the best possible resolutions do not always eliminate the underlying conflict in mapping. This conflict may be suppressed as part of the constrained variety of the system waiting for an opportunity to reassert itself. Melko describes leadership in terms of individuals with a "better idea" or an idea for a "better arrangement" who are able to convince others to agree with them and validate their new model. If they are effective as leaders, a group will form around them and their ideas. If the idea is accepted by the group but another individual is considered to be a better leader, the first one may be set aside. With a successful leader, the group may make a stronger bid for the system to accommodate their revised role expectations or role demands; this leads to role conflict which must be resolved in one way or another in order to avert transformation or disintegration of the system.

Socialization

Integration of a social system is achieved through the bonds that develop between individual psychic systems as a result of their need to relieve their own stresses (solve their own problems), the bonds between individual psychic systems and various social subsystems, and, through these latter bonds, the bonds between the individual psychic system and the inclusive system. Because subsystems come and go and have to be replaced, a social reproductive system must accompany the biotic process of producing new members of a social system. Socialization is such a reproductive system.

As various levels of animal life acquire increased capacity for learning (mapping self and world through experience), the need for organized socialization processes increases, in order to provide suitable individual subsystems and to preserve the system. In the less complex forms of life, social patterns are controlled genetically. In the more complex forms of life, individuals cannot survive without acquiring some mapping of self and world after birth. This is most notable in humans, where behavior is rarely instinctive or genetically determined. How human children understand and evaluate themselves, their society, and their world all is given to them by the social system in which they acquire their mapping (are socialized). More precisely, our understanding of ourselves and our world is the result of our interaction with the socializing processes in our social system. From the moment individuals begin the mapping process, the social system feeds them information about themselves, their world, and the results produced by various interactions.

There are many levels of interpretation and satisfaction. For example, psychology tells us that children seeking interaction learn to misbehave in order to get attention, even if it is disapproval. Similarly, adults seeking recognition may commit a crime in order to feel important. Another common example is the neurotic cycle, whereby individuals seek to relieve the stress of self-rejection by some action that in the end only increases their self-hatred and, thus, renders almost irresistible the temptation to indulge in the same action again. Such is often the case with overindulgence in food, alcohol, or drugs. There are many other examples one could choose from the unconscious workings of defense mechanisms, such as sublimation, displacement, or projection.

Learning, the mapping of self and world, is not a simple matter. Neither is socialization. One can read the laments of parents and leaders over the deviance of young people in very ancient literature. Applying a general systems model to the problem may be enlightening. This approach may not provide answers, which must come from a specific body of knowledge, but it can help to clarify the nature of the problem and set it in a broader perspective. More specifically, this approach has the advantage of removing the problem from the connotative morass of emotional interpretations and assumptions.

If we view the infant as a biopsychic subsystem in process of adapting to the social subsystem in which it is to survive, our basic question concerns the nature of the interacting systems and the nature of their interaction. As with any subsystem, children seek to satisfy their own needs and desires, and thus, above everything else, to preserve themselves, to relieve its stress through some kind of equilibration. Because the subsystem needs the system in order to relieve its stresses, successful interaction with the system must take place. Conforming to the system can thus be the direct route to equilibration for the subsystem. This is true of simple physical and biotic systems, where antithetical elements in subsystems must wait for external relations to induce changes. At more complex levels, where the subsystem (animal, for instance) has some options with reference to interaction with the environment, the problem becomes more complicated.

At every level of systems there is tension between subsystem and system. As we have seen, this tension arises because each subsystem has potential for relations that cannot be actualized in any one system—antithetical elements. At the human level, this means that no family, no group, can ever satisfy all stresses arising in any given individual. There must be a selective process, and the process by which the selectivity of a social subsystem is provided as information for the mapping of an individual psychic system is socialization. The subsystem can be expected to try to relieve its tensions the best way it can through a process of selection and rejection. The picture is further complicated by the condition that the subsystem (child) has its own subsystems that, from the beginning, make their own autonomous demands. Children cannot sleep and play at the same time. (It is possible for an infant to sleep, eat, urinate, and be held all at the same time, but such an accomplishment in later years would not produce very satisfying results.) Socialization requires selective discrimination of those stresses that can be relieved, and under which set of conditions, and those that must be constrained. Apart from the constraints of the social system, the individual must make selections simply because of time and space limitations. I cannot, for example, stay up to finish reading a book and go to bed early on the same night. When I have free time (a weekend or a vacation), I cannot do everything I might like to. More possibilities exist than can be fit into the time we have and our capacity for activity. Many of us have the experience of planning much more for any given time than can possibly be achieved, such as in repairing a car or teaching a class.

These constraints of time and space, combined with the fragmentary nature of most psychic systems, create a situation of considerable tension. As long as there is a dominant mood, feeling, or need (stress), a choice is easy: A high degree of unity is experienced under a significant stress. In the absence of dominant stress, or the presence of a mixture of strong stresses, it may be extremely difficult to select one over another, or to feel satisfied after the selection. If we have a reasonably well-developed map of our life and goals in their various aspects and of their relations to the world, we find it much easier to make difficult choices, correlating possibilities with the possible development of the map. Such an integrative grasp of our own psychic system and such mapping of its present situation and goals are usually the accomplishments of our later years. Children cannot resolve their inner conflicts in this manner. However, the social system does have a controlling superstructure that provides limits or constraints against which choices can or must be made. Socialization is the process through which a psychic system learns to order its own inner fragments and its selection of stresses to be relieved or constrained so as to enhance its interaction with the social subsystem and thus enhance its chances for satisfactions. The stresses pressing for relief do not necessarily remain passive under constraint, which often results in impulsive, rebellious behavior. When a psychic subsystem, due to circumstances, becomes temporarily dominant—demonizes the whole psychic system—concern for interaction with the social subsystem is cast aside, and the psychic fragments' demands for satisfaction become total. In children, this often results in what is commonly called a temper tantrum. In adults anything from a violent explosion of temper or a wild drunken spree to a psychotic episode could occur. As children develop, they require strong controls to keep order in their selective processes. Depth psychologists offer a good explanation for the fear children develop concerning their own uncontrolled psychic system. Children, they point out, fear that acting to relieve the wrong stress, or to relieve the right stress at the wrong time, will result in loss of the controlling social subsystem (family or parent). Emotionally, of course, this means alienation and utter chaos. Thus socialization is not merely a process through which individuals learn to behave or to adopt a cultural model. Rather it involves the dynamic correlation of psychic subsystems with the psychic system as a whole, and with various social subsystems. It does this with the emotional coloration of a life-and-death struggle to find identity, security, and relief from stress.

If society presented a single consistent pattern as the context for a developing psychic system, one would expect the development to be, for the most part, direct, simple, and successful. Margaret Mead's Samoa and New Guinea seemed to approach this kind of stability of social structure and role relations. The situation is different, and more difficult, in a complex, pluralistic social system. While Margaret Mead found nothing comparable to what we call adolescence in Samoa, our young people suffer what we call identity crises. The difference concerns the nature of the socialization process in a simple, as contrasted with in a complex, pluralistic society. In our social system, young people learn to recognize themselves in many different and conflicting roles, roles in which attitudes are often hostile and judgmental toward each other. They have ample opportunity to learn that many individuals regularly live and behave in ways children have mapped as forbidden. They may also learn that what they have mapped for their identity, meaning, and security is nothing more than the pattern of their social system at that particular time and is no more right or wrong than many others.

We need more information concerning the compromises and adjustments made in psychic systems in order to accommodate the impact of pluralistic relativism. We know something of the mind's ability to compartmentalize, that is, to keep conflicting subsystem fragments separate. Depth psychology provides us with a long list of defense mechanisms through which the psychic system keeps up the appearance of integrity, order, and conformity. If a manageable set of roles can be maintained without serious conflict, a psychic system can believe wholly in whichever one arises to its correlate social subsystem.

There is, however, another possibility, finding a sense of identity that transcends all specific roles and social systems. All of the great religions of the world have presented methods for attaining such a transcendent sense of meaning and identity, and some individuals seem to have succeeded in attaining such a perspective. However, social systems soon organized the teachings of the founders into standardized roles and lifestyles. What began as a path to a transcendent sense of identity soon became one of society's institutions. However, the significance of these individuals is not measured by their success in moving the masses, but in the very fact that they came into existence. They, together with the emergence of the scientific method, indicate that a psychic system is capable of developing the ability to view itself with a high degree of objectivity and detachment. This possibility is attested to in the ecstatic writings of mystics, in the canons describing scientific objectivity, and in the nausea and despair described by existentialists. The upper level of socialization in a complex society appears to be the emancipation of the psychic system from a narrow belief in, or identification with the mapping it has acquired through the process.

Here, too, there is a developmental continuum, beginning with the infants seeking controlling identification of themselves with their social systems, and ending with a transcending perspective, a sense of identity with the universe, a god, or nothingness, and a relational evaluation of everything. Proper understanding of socialization sheds some light on the way the process has evolved as social systems have become more and more complex and the way these changes have affected the results of the process in the patterns of development in psychic systems.

Organization

There are alternatives to the organization of anything in the universe. The universal process appears to be continual motion from one organization (system) to another, through variations, and back again. Changes seem to take place as the equilibration processes throughout the universe interact with one another and produce various compromises among the competing forces. For humans, this changing of patterns becomes a matter of creative imagination and, to some extent, of choice. When Eve saw that the fruit of the forbidden tree was desirable, she merely perceived that the arrangement after eating the fruit would be more desirable than existing circumstances "in Paradise"; that is, knowing alternatives and making choices is preferable to innocence. It is interesting that Adam and Eve are represented as having a choice about choosing or about being aware of alternatives, for it is a fascinating bit of naivete to suggest that man could be aware of a choice about becoming aware.

Man has always tended to believe that there was a glorious past and a glorious future between which the present misery must be explained. Most often, the explanation has been that man spoiled it all by his passions and greed, and the glory must be regained by some means or other, usually in another life or condition. We do spoil things for ourselves, and we are our own worst enemies, but that is merely to say that our creative solutions to problems do not always work to our advantage. The ideals of humane and mutually beneficial social relations have been clearly stated and insightfully described for at least 5,000–7,000 years. These ethical principles are, from one point of view, statements of good system principles; that is, one subsystem ought not harm another, and all subsystems and ought to be oriented toward the good of the whole system. There is always a mutual dependence between a subsystem and its system or a system and its supersystem, for no system (except the universe) exists in total independence, but there is no perfect or final arrangement, no system that would provide the "best" for all levels of subsystems. Every actual arrangement necessarily negates possibilities that would benefit some levels of subsystems, and provides less than the theoretical potential for some subsystems.

There are polar limits within which arrangements must be maintained if systems and subsystems are not to disintegrate. If subsystems lose autonomy to a degree that prevents them from controlling their own internal relations, they cease to be identifiable as subsystems, and their sub-subsystems relate directly to the inclusive system. This has been the fate of some county governments as cities and towns begin to relate more and more directly with state government. At the other extreme, if a system loses its dominance to a degree that prevents it from maintaining control over its internal relations, then it will disintegrate. Something like this followed the death of Alexander the Great. It happened repeatedly in China whenever an empire disintegrated into territories controlled by competing warlords.

The polar extremes are total system dominance or total subsystem autonomy. Either one will destroy the system. Successful integration is always between these extremes, and represents some kind of equilibrating compromise. The insoluble ethical, social, and political problem is that there are many variations possible on the continuum between these extremes. No ethical principles, no matter how noble, can describe the ideal resolution. If all agreed that everyone should be fair, the percentages of the gross national product that ought to go to labor or management would still remain unspecified. The standard of living of American workers is much higher than it was; but the more it rises, the greater are their expectations. Is this right or good? It is only clearly bad when it hurts the system—and eventually the workers. They need a healthy system in order for the quality of their share to improve. How large a share can they take without crippling the capital investments that make the system possible? The same problem exists for management, and for every other subsystem.

A government or culture, consciously or unconsciously, settles for what evolves as a workable solution. If it works long enough and well enough for most subsystems to make the necessary accommodations, it becomes a "working standard" for proper behavior. Then it becomes the "norm" in relation to which other possibilities are resisted and rejected. The more successful the organization (bonding or interrelation of subsystems), the greater the inertia of the system. Those subsystems that have made a satisfactory adjustment resist change. Those subsystems that, in Melko's terms, prefer another arrangement and perceive a problem, constitute some of the constrained variety within the system. As with all systems, social systems exist in an equilibrating tension among polar forces. What Jung described as the "shadow side" is always there, waiting for a social crisis to provide the opportunity for a demonic takeover or to force another accommodation of the system.

Demonization is the attempt by a subsystem to totally dominate the whole system, to take the whole system into itself. This is clearly impossible because everything else would cease to exist and the subsystem would have to become the whole universe. In theological language, this is the "will to be God" and is dramatically presented in the story of the fall of Lucifer. Actual demonization is the attempt of a subsystem to move toward total domination and, unless it is contained in some way, it inevitably leads to some levels of disintegration. Organization is a social system's protection against demonization by the whims or passions of disaffected subsystems.

Organization, too, can be a form of demonization. Where the control system falls into the hands of a fanatical minority that will go to almost any extreme to maintain its chosen pattern, demonization occurs. Demonization is still the domination of the whole by a fragment for the purpose of making the system serve the subsystem, but in this situation, the forces of organization are working for it as well. History describes an abundance of tyrannical social systems. Within such systems, most development and differentiation must go underground and become powerful antithetical elements waiting to disrupt or destroy the prevailing pattern.

All of this is a matter of degree and perspective. One man's demon is another man's saint. It is easier to understand and cope with the differences in degree and perspective if all social systems are seen as compromises among system levels between the polar extremes of subsystem autonomy and system dominance and in relation to the degree of common mapping among various subsystem levels and the pattern of the inclusive system. The more complete the common mapping, the more dominance is seen as participation. The weaker the common mapping, the more dominance and autonomy are seen as demonic. The goal of subsystems mapped for easy conformity to the pattern of the inclusive system has been viewed by many as the ideal society. It is the nature of the earthly paradise in which every hive of bees lives and of the state to which Adolf Hitler aspired for the whole of humanity. What can disrupt this ideal social system is the human psychic system and its capacity for differentiation, specialization, generalization, and high level symbolic systems.

However, the individual psychic system is more potential than actual until it interacts within a social system. "Socialization" is not just a process for making new citizens conformable to the social system, but an essential process for developing individual psychic systems to the level of usefulness within the social system. In human social systems, this includes some degree of specialization, even if no more than the different roles assigned to the two sexes. The more the society develops, the more specialization and differentiation of function is required.

Melko's disturbing insistence that the engine of social system movement is "problem creation" is a helpful perspective. One usually thinks of motivation, whether individual or social, as problem solving. This appears to be true of primitive societies, which change slowly over much time by making adjustments to environmental conditons or to some occasional internal problems such as overpopulation. Melko, however, was suggesting an explanation for the difference that appears in civilized societies, a pattern of growth and development toward complexity that external relations cannot fully explain.

We are all familiar with statements like, "The more we know, the more we know we do not know"; every time we solve one problem we become aware of several more that our solution has either caused or brought to light. These claims are in good agreement with a general systems perspective with regard both to increasing differentiation and specialization of information and to the accommodations required by specialization and differentiation in social subsystems. Increasing detail of information brings with it an awareness of the greater detail that is needed. When we find new data to solve a problem, we find we need more data to explain the solution. We have abundant evidence from urban-renewal efforts and from our present ecological crisis that specialized solutions bring with them a complex series of problems; individuals seeking psychiatric help to solve their personal problems will probably find that many of their former relations require adjusting and that some may no longer be suitable; and the convenience and savings provided by supermarkets and shopping centers have made the automobile a necessity and contributed to the problems of pollution and an energy shortage.

The relation of the process of problem creation to the dynamics of a social system is this: Problems are created because they are perceived as problems, and if they are responded to, there is further specialization of information and social subsystems. The question of the possible response brings us back to organization. Some problems and their proposed solutions will fall within lines of development that fit comfortably with existing structures; that is, the line of development does not appear to alter the existing nature of system–subsystem relation and balance. (We use the term "appear" because what will actually be may not be evident until much later.) Other problems and their proposed solutions will appear antithetical to existing solutions and, if possible, be constrained. The "establishment" may deny that there is any problem and maintain that no solution is needed or merely claim that the proposed solution is worse than the problem. The system cannot possibly handle all the problems perceived by all levels of sub- and sub-subsystems. Problem perception or creation grows along with specialization geometrically like a cancer or an atomic reaction and would lead to the disintegration of the system if the process were not controlled. The process of selecting some specializations and gradually incorporating them into the system is the normal process of successful organization. Polar tension and constant equilibration must exist between developmental specialization and organization or integration.

Again, there is a continuum between the extremes of strong bonding and inertia to the seeming chaos of a period of cultural transition. The attempt, perhaps begun by Aristotle, to describe the different kinds of government overlooks the degree of difference between one situation and another. Classification can be useful here, as in all fields of information, but there is always the danger of losing sight of the degrees of variation seen in comparing one form of organization to another.

We have referred, throughout this chapter, to the inclusive system without regard for the external relations of that system. The system might be a tribe, a feudal city, a nation, state, or an empire, but no social system exists in a vacuum. The external relations of an inclusive system have a significant effect on its internal relations. The relations of subsystems to system are themselves dependent on the system's relations to other systems, to the supersystems in which it participates, and ultimately, ecologically, to the whole universe. To incorporate this added perspective, it is necessary to turn our attention to the problems of macrosocial systems.

Chapter 11

Macrosocial Systems

Size, density, and complexity are all directly related to significant differences among systems. Each step of the way, from electromagnetic energy up through each new level of systems, involves some increase in size, density, and complexity. Molecules contain structures and processes that are arrangements of the already complex organizations of particles and atoms. The same principles hold for social systems. As social systems increase in size, they either divide (mitosis) and form new units, such as a family or village, or they grow and develop in density, complexity, and size. In the latter case, the management of these increases necessitates the development of more elaborate internal relations, which, in turn, produce differences in external relations.

The similarities and differences between simple and complex social systems constitute a fascinating field of inquiry in which there are still some healthy differences of opinion. However, on some characteristics there appears to be considerable agreement. Once more, we must rely on the writings of the experts and attempt to set their conclusions into the perspective of general systems theory. Fortunately, Melko's *The Nature of Civilizations*[1] offers an outline of what appears to be a generally acceptable model. A comparison of this with *The Evolution of Civilizations* by Carroll Quigley, *The Primitive World* by Robert Redfield, and *The Origin of Civilized Societies* by Ruston Coulburn[2] provides a body of interpretation that fits naturally and easily into a general systems perspective.

---

[1] Melko, Matthew, 1969. *The Nature of Civilizations*. Boston: Porter Sargent.

[2] Quigley, Caroll, 1961. *The Evolution of Civilizations*. New York: MacMillan.

Redfield, Robert, 1969. *The Primitive World and Its Transformations*. Ithaca, New York: Cornell University Press (originally published 1953).

Coulburn, Rushton, 1959. *The Origin of Civilized Societies*. Princeton, New Jersey: Princeton University Press.

What has just been referred to as simple and complex social systems are traditionally called primitive societies and civilizations. Any terminology has its weaknesses, and both will be used here to keep the relation clear. That of general systems theory is generally applicable to all levels of systems; the conventional terminology has the advantage of referring to a specific recognizable class of entities. Using both is necessary, just as accurate knowledge requires continual dialectical movement from the general to the particular and back.

Some Central Problems

One of the central issues considered in these books is whether there is a difference in kind between primitive societies and civilizations. Another central problem is whether a civilization is a real entity or just an intellectual convenience created in the minds and writings of scholars. A third fascinating problem is whether there is a general developmental pattern applicable to all complex social systems. A fourth problem is whether, granted that civilizations do develop, there is any regularity, necessity, or ordered sequence in their development. A fifth problem, really a family of problems, is whether a certain social configuration is (or was) actually a civilization, that is, where to draw the line between real civilizations and those that only resemble them.

A specific answer to the last question depends on the way the preceding problems are resolved. Assuming that one can describe the relations and boundaries that must exist for an actual entity to be called a complex social system or civilization and can distinguish it from other kinds or levels of social systems, one must still resolve the problem of social systems that do not quite fit the full description. In some ways, this is part of a universal taxonomic problem, but it can be a very instructive area of investigation. In the case of complex social systems, the entities under consideration are so large and observable that the degrees of system formation can be seen rather clearly. They have been described in many scholarly works and can be used as a model for understanding systems that are more difficult to observe and about which there is less information. Once more, the suggestion here will be that the differences are of degree. Civilizations, like all other systems, begin to exist in degrees, develop in degrees, and disintegrate in degrees.

Application of a general systems model to these problems requires caution. It is essential in any area of knowledge to define or describe the entities of concern. Human knowledge has always been "haunted" by supposed entities that were "felt" or "known" in some way but were beyond analysis. Ghosts, spirits, demons, souls, and gods all have the advantages and disadvantages of being beyond analysis. That is, one knows nothing of their internal relations and little of their external relations. Their behavior patterns are seldom repeatable or verifiable except in special circumstances. If one accepts that to be is to be related, there is no doubt that these entities exist in relation to some psychic systems, but the question of their relation to other parts of the universal system remains in doubt. In the history of human thought, there are many examples of the tyranny of words or terms over observation. Our purpose is to use the eyes of experts to see as clearly as possible what there is and to describe the way it conforms to and correlates with the universal system of systems. It would be surprising if what we found fit neatly into any pattern of abstract classification. From what has been described thus far, one should expect to find processes of formation, development, transformation, and disintegration, with one level of complexity gradually evolving into another.

Primitive (Simple Macro-) Social Systems

In the preceding chapter, it was mentioned that individuals are constantly forming relations of a casual sort. Some are maintained over longer or shorter periods of time, and some are relatively permanent. Writers in the field make distinctions among these relations by using a series of terms indicating degrees of organization. Quigley distinguishes among collections, which are those temporary social groups that tend to be the most lasting and usually appear as subsystems within a society and societies. A society is identifiable because it has a culture and organization of its own that are the foundations for a sense of autonomy and identity for the social system and the context of meaning within which social subsystems and individuals define themselves. In *Human Patterns*, Melko suggests further differentiation.

In *The Primitive World*, Redfield gives a clear and generally accepted description of the nature and transformation of simple macrosocial systems. The internal relations constituting organization and processes are relatively simple and similar around the world. Hierarchies tend to be shallow and broad; they contain few levels, with most subsystems only slightly differentiated or specialized. There is usually a chief, a shaman, and perhaps a council of elders or its equivalent. Often, there are clan subsystems or work (function) subsystems that organize families as their subsystems. Families, in their turn, organize individuals as their subsystems. Work or functions may be divided along clan lines, between the sexes, among age groups, or varying combinations of all of these. Whatever its particular form, the hierarchy is generally well integrated, with relatively little fragmentation or overlapping.

The control system is general and informal (not written or codified) and based on oral tradition. Common mapping is thorough. The society has a common moral view of the meaning of the world and the place of the social system within it. In this context, the word "moral" refers primarily to volition, the natural universe is understood to be governed by decisions and choices rather than impersonal laws and processes. An important aspect of the control system is that the individual and social subsystems have meaning and purpose only in relation to the culture's moral view of the world, that is, the common mapping within the social system. The control system functions through ritual and institutionalized roles. In this case, "common mapping" means that the individuals' mapping (understanding) of themselves is derived from participation in the psychosocial drama through which the social system defines and organizes itself. This is true in degrees in any social system but, in the case of simple macrosocial systems, the limited amount of constrained variety means that there are few alternative role relations. The Garden of Eden model is a reminder that there are always alternatives (antithetical elements), but the long term stability and consistency of primitive societies provides ample evidence that the potential for variety is low, subsystem autonomy limited, and the control systems very effective. In this connection, the word "participation" implies a process of mutual definition among the system and subsystems in which each contributes to the definition of the other and the whole. Put differently, there is a high level of correlation between the integration of individual psychic subsystems, social subsystems, and the inclusive social system.

Information is stored primarily in individual psychic systems, and the communication system is primarily oral in casual conversations and ritual reinforcements. Constraints are built into the system because the meaning of subsystems includes participation in the common mapping and relations of the system. In other words, the alternatives to conformity are alienation, meaninglessness, and anxiety.

Peer relations are primarily cooperative and highly interdependent. Competition and autonomy function in relation to fairly simple matters because of the lack of opportunity for any significant variety. Such variety as may emerge in imagination or fantasy is handled through institutionalized understandings (common mappings) of the role of spirits and demons, both good and evil. Peer relations are usually regulated through the organization of the social subsystems that have set the role relations for which the young members of the system are mapped (conditioned, trained, etc.).

There is always some degree of autonomy for any subsystem since without it the subsystem would cease to exist, and because the basic subsystems of a social system are usually considered to be the individuals in that society, individuals always have some degree of autonomy. This may be clearer if we consider that the basic unit of a social system is a particular and unique genetic code that emerges in a fertilized egg. It is the interaction of that code with input from the environment that determines the development of the individual. Human individuals are adaptive-process systems that evolve as input from the environment is processed, responses are developed, information is adjusted and stored, subsystems are developed and organized, and control systems are correlated with external relations. The potential for variety in the original genetic code is immense, as can be seen in the variety that has developed in complex pluralistic socieities, as well as the variations from society to society. However, because the psychic systems of individuals develop only in interaction with a psychosocial subsystem or systems, the possible variety is strictly limited to the kinds of interaction available between a particular genetic system and a particular psychosocial system. Because each genetic code is unique, there will be individual differences in style and manner, regardless of the strength of system controls. The extent of these differences (allowable variety) is a matter of degree and will always lie somewhere between the nonexistent poles of total autonomy and total control. In simple macrosocial systems, the level of autonomy and variety is such that it does not usually disturb the organization and processes of the social system. Because of the simple structure and organization, homogeneous information storage, and common mapping processes, differentiation occurs largely at the personal or class level, rather than reflecting nonintegrated variety within the system.

Where there are existing systems, there must be some degree of competition, just as there is some degree of autonomy. In simple, well-integrated systems, competition is largely limited to personal or social subsystem advantage in position within the hierarchy, that is, in relation to sources of power and privilege. Such alterations do not usually change the basic process–structure pattern of the social system, even if the constituent subsystems change. Social systems differ from organic systems in a number of ways, including the fact that social subsystems can exchange functions without being destroyed in the process. In a simple system, the pattern of system organization remains very much the same through such changes—the new ruling individual, family, or clan rules over the same basic system. However, owing to the general conditions described above, cooperation among peer subsystems and within hierarchical relations usually predominate.

The amount of constrained variety is, as already mentioned, limited. However, one must not overlook the known psychological facts concerning perceptual interpretations or the large body of evidence illustrating selective perception, selective memory, and the distortions of psychological defense mechanisms. On the other hand, one must keep in mind a similar body of experimental data demonstrating the considerable power of social validation. In the latter case, individuals "see" what the group "sees" even if it is contrary to their own senses. In a social system where patterns of behavior are simple and universally reinforced, the pressures for conformity are largely irresistible. Even the variety that is genetically and environmentally unavoidable is controlled (suppressed or repressed) by the dominance of common mapping.

As suggested above, information storage for the system would be within the individual psychic systems of individuals, primarily those, such as leaders and elders, who have a specialized function for information retrieval (recalling tradition). This basic information is supplemented and enhanced through ritual actions and patterns of manners and morals.

No social system can exist without some degree of specialization, because the beginnings of basic specialized functions and relations are biological.

Some degree of special mapping and information, for example, separates men from women, and both from children. The chief, shaman, or clan leader has more privileged mapping and special perception. Shepherds, farmers, and hunters have special information that may not be shared with the populace in general. These elements of specialization, however, follow the nature of elaborations or developments of the basic patterns of common mapping, and are not likely to contribute to any antithetical development.

Even though the potential for antithetical development is always present, integration is sufficiently thorough to limit fragmentation to an occasional maladjusted individual or an unusual rivalry among individuals or social subsystems. Most tensions of these kinds are relieved (equilibrated) through institutionalized processes (rituals), and the system as a whole continues largely unchanged. For a subsystem to exist—that is, to be recognizable or describable as a separate entity—it must have a recognizable degree of autonomy.

In at least one instance, two opposite forces cooperate to increase the inertia of the system. Subsystems normally tend to become institutionalized, that is, to become conformed to functioning in the service of the system. They also tend to become part of the pattern and identity of the system. At the same time, each system or subsystem tends to preserve itself as long as it can. Because the subsystem that has been institutionalized "re-cognizes" itself as a part of the system, it continues to maintain itself as such, even when its function is no longer needed. Hence, we find social subsystems, parts of information systems, and process systems that are preserved for which there is no current use. In fact, it is very difficult to eliminate such remnants, because the whole system tends to rally to sustain them. In primitive societies, one often finds bits of ritual for which there is no longer an explanation, and aspects of social organization for which there is no apparent use. This tendency becomes more pronounced in larger, more complex social systems.

System Autonomy

The question of the relative autonomy of a system is one that resonates up and down through every level of existence. Every system is a subsystem in a larger context, and every subsystem a system internally. Thus, what we may discuss as system dominance from one perspective becomes an analysis of subsystem autonomy in a larger context. There is, however, a wide range of differences in the degree of participation from one level to another. Every system is a subsystem in some larger context, but not every system is a subsystem to the same degree. For example, a son of nine years is a subsystem of his family and is best understood as such. While the same son, at age 50, is still a subsystem of the aged parents he visits, it is not as useful to consider him as such. The difference is in the nature of the bonding, mapping, and control of subsystems and the way they interrelate. Children relate to their families through tight bonds, strong controls, and in early childhood, extensive common mapping. Nine or ten years of age is often a kind of golden age of "subsystemhood" for a child. Thereafter, the variety of mappings, conflicting bonding and control subsystems, antithetical elements, and pressure for increased autonomy accelerate the processes of disintegration.

As a system increases its autonomy, it also increases the processes that are determined by internal control systems and decreases the external bonds that connect it to the internal processes of an inclusive system. Some systems have a high level of system autonomy because the inclusive systems in which they participate are either very simple with few, if any, controls, or else the controls are accidental and indirect. Consider a diatom of some element or a simple molecule in outer space. There is no doubt that it is a system and a subsystem of the system of cosmic dust. However, the inner molecular system is highly autonomous under these conditions and may be unaffected by its inclusive system for long periods of time. Similarly, a primitive tribe is a subsystem of the ecosystem in which it participates, and yet its processes as a social system may be unaffected by the ecosystem for long periods of time. Such a primitive society, as a social system, has a high level of autonomy because there is no inclusive social system. Again, we return to the basic notion that things exist in the degree to which they relate to other entities. Such a primitive society is also highly dependent on the ecosystem in terms of energy sources and freedom from destructive forces. However, as a social system—in terms of human social relations—it is largely autonomous because it does not receive mapping or controls from a larger social system. This agrees with Melko's definition of a society as a social group that has its own culture. If by culture, Melko means the internal social values, goals, and inhibitions comprising a control system, together with matters of style, aesthetic preference, and religious belief, then we are both saying the same thing. The concept of civilization is needed to indicate that this autonomy has moved beyond the limits of primitive societies to that characteristic of larger social systems. This does not mean that the included primitive societies have been totally and immediately homogenized. It means only that some controls, some mapping, and some bonding have moved to a more inclusive system. We are all aware of some new, legally constituted nations that have relatively few internal bonds and little integration. When the bonding, mapping, and controls move beyond nations, empires emerge as the inclusive system. An empire is a social system to some degree that is to be analyzed in terms of its internal relations such as common mapping, bonds, and controls in other words, common culture.

The term "civilization" seems to refer to a social system that possesses the degree of autonomy just described and whose processes are developmental rather than steady. Nations or primitive societies may be held under certain military controls and called an empire but do not actually constitute a civilization until some significant level of integration has taken place. As is true for atoms, molecules, and every other system in the universe, inclusive systems form when systems are held in proximity long enough to form bonds that begin to shift some aspects of system autonomy to the emerging supersystem. The point at which a supersystem is said to exist is a matter of convenience. A significant description would explain in which aspects the supersystem exists. There is no value in debating the existence of a certain civilization. It would be much more useful to indicate the number of bonds and controls that have or have not emerged, to emphasize the formation process instead of struggling with arbitrary problems of taxonomy.

Thus, strong internal controls do not necessarily indicate a high level of system autonomy. If the pattern of dominance derives from and correlates with participation in the mapping of a dominant supersystem, then there may be little relative autonomy at either level. In this case, there is close correlation and integration of the information and control systems of each level. An example would be an authoritarian family in an authoritarian social system. On the other hand, a pattern of strong system dominance may arise in a situation where there is little relative autonomy for subsystems internally but strong autonomy for the system relative to external relations. This would be the case for an authoritarian nation-state or tribe with few significant "international relations."

Subsystem Autonomy

The principal difference between system autonomy and subsystem autonomy is that in the latter case dominance from a higher or supersystem is considered signficant in the description of a given system level. This is an important difference, because designation as a subsystem indicates a tension or set of tensions that the term "system" ignores. Because integration is never total, and is in many cases fragmentary, a study of the balance of system integration and subsystem autonomy is essential to the accurate description of any system. Of equal importance is a careful analysis of the nonintegrated aspects of subsystems. The prefix "sub" if appropriate, indicates some degree of integration as well as some degree of nonintegrated or constrained variety in the composition of the systems to which it is applied. It is the presence of nonintegrated variety that supports the perception of problems and creative solutions. Melko makes the point that stresses often exist that are not recognized as problems because no alternative appears to be possible. The more constrained variety (antithetical elements) exists within subsystems, the more likely it is that alternatives will be perceived.

Sources of Variety

Variety in subsystems appears to increase in social subsystems through at least three processes. The first is an integrative process, through which differentiated systems are bonded into a higher level system. RNA molecules replicate themselves by attracting the appropriate proteins, sugars, and other molecules and by holding them in order until they bond into a new macromolecule and are released. In this case, a degree of constrained variety is present at every level from subatomic particles up through atoms and molecules because, at each level, other bonding patterns are possible. However, the simple, stable nature of the systems and the strength of the integrative forces of the inclusive system are such that the variety is likely to remain constrained unless acted upon by some strong external force. From the level of the largest macromolecules on up, the presence of constrained variety in systems is most in evidence through the systems' tendency to disintegrate. From the level of cells on up through social systems, the need for constant maintenance and reproduction of parts indicates the difficulty with which the variety is constrained. The principle that the more complex the system the more energy is consumed in its preservation testifies to the burden of increasing amounts of constrained variety.

Whenever existing systems are integrated into an inclusive system, variety is brought into the system, and the process of integration requires that all variety antithetical to the emerging subsystems of control, communication, and cooperative processes be constrained to some degree. New social systems, such as a mining town, an empire, or a pair of newlyweds, provide interesting examples. The strength of the integrative process depends on the number and strength of the system aspects participating in the new supersystem, as compared with the number and strength of the aspects that are judged to be antithetical and in need of constraint. This is clearly a case of systems yielding some degree of autonomy and becoming subsystems. The degree of subsystem autonomy these systems preserve is in direct relation to the degree of system dominance the supersystem can achieve. The American Continental Congress, the U.S.S.R., and the United Nations all represent varying degrees of supersystem emergence. In each case, variety is imported in the process of bringing the supersystem into being. Many who write about the emergence of civilizations think in these terms. They argue that representatives of a variety of differentiated social systems were drawn together in some fertile valley, after which, as Quigley suggests, there followed a period of mixture and a period of gestation. This seems likely, but it is not the only possibility.

A second way in which variety is produced is through the differentiation and specialization of subsystems. Variety indeed emerges in this way in developing civilizations. Quigley gives an interesting account of how processes of production, invention, and control combine to encourage growth and, inevitably, variety through differentiation and specialized function. However, much concern has been expressed as to the beginnings of development (civilization) in simple primitive societies. In other words, how and why does development become systemic and self-perpetuating?

Biologically, differentiation and specialization proceed according to a preestablished genetic pattern until a certain stage of maturation is achieved. In social systems, there is no genetic pattern, only the common mapping of the system and the mapping process applied to new members. However, psychic systems are capable of great variety, and mapping processes must provide for "proper" development and inhibit "deviation." Each individual is biologically unique, and the interaction of that individual with a social system produces a unique psychic system. This would produce continual variety, were it not for the great importance of the social system. Human beings need social validation; that is, human systems have a strong tendency to bond with other human systems. At various times, there is new emphasis on individuality, and conformists are condemned.

The social systems must then invest considerable energy in the processes designed to keep the individuality (variety) of the component psychic systems from destroying the system. Of course, social control systems can become oppressive, and the cry for subsystem autonomy (individual freedom) is a normal arc on the helix of social change.[3] The social system can be successful in mapping subsystems because human psychic systems actively seek and maintain bonds with peer systems within it. However, as noted above, no integrative process can be complete without destroying the component systems. Hence, constrained variety is present in some degree in subsystems of every level. The presence of variety is the possibility of invention. The right combination of stress (crisis) and applicable variety can produce novel solutions. The stress need not be starvation or war. It need be no more than the awareness of a difficult or painful aspect of life that has been transformed into a recognizable problem by the imaginative invention of an alternative arrangement. Overpopulation of a fertile river valley is one example of such a problem.

The potential for some development is always present if there is sufficient variety of constrained mapping among subsystems. The potential for development depends on the constrained variety producing the perception of problems and the readiness of some level of social organization to accept the alternative arrangements of a proposed solution. Once a subsystem adopts an alternative arrangement, it acquires an added degree of differentiation. If these novel aspects are then related functionally to the internal processes of an inclusive system, the subsystem differentiation becomes a specialization within that system.

The significance of this distinction lies in the fact that specialization requires specialized mapping. Put simply, subsystems with a specialized function see things differently. In ancient civilizations, the priests and warriors often disagreed. This is usually analyzed as a power struggle between

---

[3]I suggest that this is a more accurate and timely metaphor than the old pendulum image, because systems never return to exactly the same condition. Situations are similar, even isomorphic, but never the same.

vested interests, but it may be more revealing that these members of specialized subsystems had different maps of themselves and their worlds, containing significant degrees of constrained variety. Moving from two such subsystems to many with vastly different specialized functions, one can see that an ever larger number of different perspectives are associated with the specialized information possessed by each of these subsystems. However, as noted in earlier chapters, the mapping of psychic systems is fragmentary, so that it and the mapping of social subsystems can be successfully compartmentalized. This means that highly differentiated, specialized mappings can develop in isolation from the integrative mapping that binds the individual or group to the society. The more complex the system, in terms of differentiation, specialized function, and depth of hierarchy, the more constrained or compartmentalized variety will exist throughout it.

Once a social system begins to develop a certain level of specialized functions and specialized mappings, it is more and more likely to have subsystems that perceive problems or stresses (i.e., to conceive of more desirable alternatives). This means that system development is a source of variety, and variety is, in turn, a source of system development. Presently, we shall consider the critical difference between a social system that changes in reaction to occasional stresses and one in which change is called development because it has become systemic.

A third way in which variety may be produced in a social system is through the reception of information from separate and highly differentiated systems or subsystems. When systems that have developed significantly different information systems—different cultures, for example—come into contact through the exchange of information, the amount of variety may increase in one or both systems. The change may involve the assimilation of information into existing information by extending or enlarging existing patterns of mapping. The information may be received by a subsystem that becomes more differentiated within its system through accommodation to the new information. In our own time, social sysems are bombarded with information from every part of the world and, consequently, there is much variety available and change is often very rapid.

In all these cases, the principle of moderate novelty applies: If the information is too strange, it is received as noise, as nonsense, and if it is too similar, it will be categorized as correlating with existing mapping. Variety arises in connection with information that is sufficiently different to indicate the need for change (a different arrangement) but similar enough not to require a whole new meaning context. I shall later make further reference to the worldwide system of instant communication, where through all these means, the potential for variety, and constrained variety, is very high.

The most critical question for understanding the development of social systems concerns the availability of alternative solutions, that is, the ability to perceive alternative sets of relations and processes. One usually thinks of

Sources of Variety 187

stress arising before an alternative arrangement is perceived as a solution. However, there are situations in which an alternative arrangement is perceived first and the perception of a problem or stress follows as a consequence. The warnings in the stories of the Garden of Eden and Pandora's Box are against not the recognition of stress but the recognition of alternatives. It is the latter that makes individuals discontented with their roles and gives rise to inordinate curiosity or ambition, inordinate because it does not fit within the existing pattern of relations and threatens to alter the balance of system dominance and subsystem autonomy. One of the most significant differences between simple macrosocial systems and complex developing macrosocial systems is the presence of variety and plurality of the mapping of subsystems at various levels. Variety in mapping implies a variety of perspectives, which, in turn, implies potential alternative solutions. The presence of variety is the essential ingredient for one kind of problem creation; that is, the awareness of alternative solutions encourages the perception of stress as a problem to be solved through change or invention.

Melko's hypothesis *Human Patterns* relates directly to this suggestion. He claims that the "engine" of civilizational development is problem perception rather than problem solution. This seems to indicate that the generation of various perspectives (variety of mappings) produces the seedbed for the perception of alternative arrangements. When some of these alternative arrangements are perceived as preferable to existing sets of relations and processes, the perception emerges of a problem or a stress, a line of thinking that relates to our earlier discussion of the role of optimal stress in producing systemic change.

It has been repeatedly suggested here that systems in general do not move or change except in response to optimal stress, stress too strong to be ignored but not strong enough to be destructive or paralyzing. Such change depends on the reception of information concerning a problem or stress. That is, external relations may be present that can harm a given system, but if information concerning these relations is not received by the appropriate control subsystems, the system will behave as though the danger did not exist. There may be various kinds of information or signals available, but if they do not correspond satisfactorily to the mapping in the control subsystems, they will be unrecognized as information, received as meaningless noise, or assimilated under some routine category of information storage and referred to an institutionalized (regularized) response. There are many alternative ways of relating and functioning, but until they enter a control subsystem as the mapping of alternative behavior patterns, they do not exist for that system. Turning that thought around, one may say that the more alternatives one perceives, the more problems one is likely to perceive; the more variety is present in various levels of subsystem mapping and function, the greater potential there will be for the perception of alternatives.

An apparent contradiction must now be noted. Anthropologists have observed some primitive societies to have been open to new ideas and less inhibited than what we have agreed to call civilized populations. A simple social system is characterized by a more simple, more general, and less differentiated mapping of its internal and external relations than a more complex social system. Consequently, new information coming into a simple system concerning matters for which there is little mapping and negative indicators may not be resisted; an example is the cultural rape and destruction of Samoa. Now the contradiction disappears. We referred earlier to problem generation or change within a simple macrosocial system, whereas the anthropologists describe what happens when a "cultural virus" attacks an unprepared systems: Like any system, it either adapts or disintegrates.

Transition from Simple to Complex

One of the most interesting problems considered by civilizationists consist of two questions. What is the essential difference between primitive societies and civilization, and why did civilizations begin in certain places and not in others? In the roughly 2 million years of human evolution, it appears to have taken about 1.99 million years for the first civilizations to begin to appear. Once they began to appear, civilizations developed independently in several different locations, including China, India, Mesopotamia, Egypt, Crete, and Central and South America. On a scale of 2 million years, these beginnings are remarkably close together. Primitive societies had probably existed as steady-state systems for a large part of those 2 million years. Because social systems exist among primates, there is no reason to assume that early humans ever lived in any other pattern. This long history suggests that primitive societies (simple macrosocial systems) are highly successful and adaptive systems. However, current evidence indicates that civilizations began to appear some 7,000–10,000 years ago and have continued to the present time. One must conclude that conditions were ready for the next level of system complexity and that the conditions first came together only in certain locations. This is the same pattern biologists use to describe the emergence of the first macromolecules in the primordial ocean. In fact, it is the basic pattern of emergent novelty and levels of increasing complexity stressed throughout this book. Again, when potential subsystems come together and remain together under the right conditions, they tend to form bonds and to evolve a more complex inclusive system with novel characteristics that differ from those of simpler levels.

The central problem thus becomes the analysis of the set of conditions under which social systems experience optimal stress and resolve this stress by developing a more complex system. From the perspective of general systems theory, the question can be clearly stated in terms of energy and complexity. The more complex the system, the greater the amount of energy needed for simple maintenance, energy that must be taken from the work or function of the system. Further, the more complex the system, the greater the amount of energy needed to synthesize the system. It is not uncommon for writers in general systems theory to refer to complex systems in the universe as improbable. As we scan the heavens and wonder whether there are other worlds with complex forms of life, we must remain aware that complex biological and complex social systems are not a commonplace form of existence. The sun is a huge energy pump, pouring just the right amount of energy onto the earth under just the right conditions to cause that energy to be organized into and channeled through ever more complex layers of systems. After some 7 billion years, a living system appeared that recorded its gathered information in symbols and then used those symbols in abstraction to invent new patterns and forms. These thinking animals organized themselves into highly efficient social systems of modest size and became biologically successful and the dominant force of animal life. As the population of a given social system became too large, some sort of mitosis or spore scattering took place to equilibrate the stress of density, conflicting variety, or the like. However, about 7,000–10,000 years ago, a stress or a combination of stresses produced a set of circumstances that caused individuals and social systems to invest the extra energy necessary to create—to synthesize—more complex and inclusive social systems.

If social systems had gradually increased in size in adapting to external stresses but had continued as primarily reactive systems, the problem would not be as interesting. Rather, when social systems reached a certain level of complexity, change and growth became systemic, and, adaptation became primarily to stresses generated within the system itself. The question can now be put in another form: Why would a system continue to develop in the absence of external stress when this requires even larger investments of energy to build and maintain the system?

The study of comparative civilizations indicates that complex marcrosocial systems do tend to grow until they begin to distintegrate. There are civilizations that have muddled along at about the same level over long periods of time and others that have gone through successive stages of development and disintegration, but the question remains. There is still a significant difference between a system that operates at a certain level of complexity over millions of years (especially if our study includes the primates) and one characterized by systemic development but inactive for a few hundred years. Complex social systems also go through various stages of development, disintegration, and transformation. The significant difference lies in the presence of an internal systemic generator in complex macrosocial systems and its absence in simple macrosocial systems. The two types of system appear similar in that each is a kind of adaptive social system, but the simpler kind adjusts to stresses as they arise, whereas a complex system generates most of the stresses to which it then adjusts. This description applies to any developmental system at any level. However, most developmental systems that have been studied closely are biological and have genetic code and control subsystems. The fascinating thing about social systems is that they are the most open, adaptive, and undetermined of all systems, with possible exception of the purely symbolic and abstract. Their internally produced or recognized stresses tend to produce specialized subsystems rather than to modify the mapping and behavior of the whole.

The problem of distinguishing primitive societies from civilizations, we have seen, is that primitive societies have, at least in simple form, every characteristic by which one would like to distinguish civilizations. Further, there are social systems that are neither clearly simple (primitive) nor really complex (civilizations). Yet otherwise, the hypotheses underlying this book would be in serious trouble. The general systems model presented here suggests that one should find simple systems capable of bonding into larger systems, incomplete transition forms that are not clearly simple or complex, as well as well-developed complex systems. The latter should have novel characteristics that are not possessed by any of the subsystems but are clearly the product of the synthesis, organization, and processes of the more complex systems. Further, the more complex systems should retain the characteristics of all systems in general. We must suppose that it is the novel characteristics of the emergent complex system that have led to the distinctions that civilizationists have found so troublesome. Where there are no emergent novel characteristics, there can be but one kind of system, with varying degrees of complexity. On the other hand, if characteristics emerge in complex macrosocial systems (civilizations) that are not possessed by simple macrosocial systems (primitive societies), then one is justified in claiming that civilizations constitute a genuinely new level of system development.

There seems to be general agreement as to the characteristics by which a civilization is recognized. Organizing these characterisics is a little like looking for the center of the universe. The difference between Ptolemy and Copernicus is a matter of convenience and elegance. Quigley centers his thinking around a means of expansion and Melko selects problem perception as his hub. Both perspectives are helpful and work well, but I have another because it emphasizes the central concern of general systems theory. Following Quigley I suggest that the necessary foundation is organized productivity. He argues that simple macrosocial systems tend to be either food gathering or food producing, and only the latter kind can become developing, complex social systems, and then only under the right circumstances. This observation is a good illustration of how perceptions change and become clear when seen through a general systems model. The terms "food gathering" and "food producing" do not in themselves reveal why one and not the other is the seedbed for an emerging civilizations. Systems analysis would emphasize that the transition from "food gathering" to "food producing" is the result of a change in the mapping of the social system to a more complex level of information, specialization, control systems, external relations, and inertia. This change is only one stage on a continuum that reaches all the way to an empire or a world social system. And, at least one civilizationist would respond, Why not a universal intergalactic social system?

Necessary Sequence, Not Necessary Development

Piaget makes a good case for the claim that psychic systems develop only when and as far as circumstances require. If they are developing, there is a necessary sequence of stages through which they must pass. Levels of complexity must build in an orderly fashion, and the periodic table makes it clear that the same is true for atoms. The bonding that creates molecules, which then combine as subsystems into complex molecules, reveals the same kind of sequential necessity. However, when one reaches the level of nucleotide-control systems (DNA and RNA), there emerges the systemic novelty of predetermined genetic-control systems. It is this system characteristic that has dominated the thinking of many in other fields as they attempt to find deterministic patterns of development in higher order systems. However, at the level of more complex control systems, genetically determined control systems tend to be supplemented by information programmed from experience. The development of a psychic system is not genetically predetermined. It is the interaction of the individual psychic system with other psychic systems within the context of symbolic meaning of a social system that produces and limits the development of the individual psychic system. Social systems have no genetic controls. Their only predetermined controls appear to be the nature of the subsystems, and of the systemic processes that evolve through bonding and synthesis or development. Empirical evidence indicates that social systems must pass through a certain number of patterns if they are to develop. The necessity is in the sequence of stages.

The general processes of stress and equilibration cannot cause a system to adapt and develop without limit. We need much more research, but there seem to be basic systemic requirements that limit the variety of adaptive patterns and establish a developmental sequence. This is not some dark mystery. It is a general systems principle that simple systems must be synthesized to some level of autonomy before the next level of complex system can emerge.

Characteristics of Civilizations

We shall not regard primitive food-producing societies as civilizations, but they are the beginning from which civilizations can emerge. The movement from food gathering to food producing is an identifiable stage in the continuum of development that leads to a civilization if development continues. The distinction between primitive food-producing societies and civilizations is that the development of the former is still largely dependent on stress from external relations, whereas development in the latter becomes increasingly systemic; that is, the major source of stress is in internal relations. This kind of analysis suggests that Quigley's "means of expansion" and Melko's "problem creation" are novel characteristics of a higher order of systems, or at least a higher degree of complex development. The essential difference in the transition is that the characteristics used to describe civilizations are in part cause and in part result of a synthesis or development. Some of these characteristics represent elements of a system that must emerge from the constrained variety of the subsystems through bonding processes during a period of transition. The transition itself is an equilibration process during which new patterns develop in organization, control, and processes so that some new entities, subsystems, systems, or supersystems emerge with a sufficient level of autonomy and adequate patterns of interdependent external relations.

From this perspective, the essential element is not the characteristics that emerge, but the emergence of new entities and bonds that sustain new functions and processes: The transition to a civilized society occurs only when the essential characteristics become systemic. By "systemic," we mean that which pertains to the entire system or to the system as a whole. Any one or several of these characteristics may appear as antithetical or nonintegrated elements in subsystems or may be tolerated variety within a developing system. When they become systemic, they have taken hold in the control, information, and energy subsystems and have been synthesized into the modified mapping of the whole system.

This is not an all-or-nothing situation. Scholars in the field can expect to find degrees of gradual change, but, as with boiling water, there is a point beyond which there can be no doubt. This is a critically important point, because development in any system necessarily requires control subsystems that equilibrate the differentiation, synthesis, specialization, and integration of emerging and increasing subsystems, novel processes, necessary functions, input, storage, distribution, and output for the system as a whole.

As degrees of development emerge and complexity increases by stages, certain characteristics appear. In what follows, information that appears to have common support among civilizationists has been arranged to fit the general systems model.

Subsystem Multiplication

One of the most clearly noticeable characterisics of complex macrosocial systems is an increased density that results from subsystem reproduction and differentiation. This is true at all levels of subsystems, from individuals and small groups up through the various levels and sizes of subsystems devoted to maintainence, production, control, or information processing and storage. The production of increasing numbers of subsystems requires some hierarchicial organization and overlapping, layered control subsystems. This kind of development requires more energy and greater density and produces urban centers. These large centers of population, of course, produce many stresses (problems) for which solutions (equilibration) must be sought. These solutions usually involve new processes requiring new subsystems, new control subsystems, and new information subsystems and increasing variety (the plurality of perspectives), which again produce stresses that press for equilibration. Clearly, with regard to subsystem multiplication, the engine of systemic development is running.

Specialization

From the perspective of the polarity of subsystem autonomy and system dominance, every system is special in some degree. That is, everything that exists has some unique relations with other and inclusive systems, even if they are no more than location in time and space. However, we generally use the term "specialization" to refer to relations that are functions of the inclusive system. Subsystems participate in systems through relations that relieve stresses (satisfy needs) in both directions. The subsystem provides for some needs of the system and the system provides for some needs of the subsystem. When a subsystem becomes the special source for relief of certain kinds of stresses, it is generally considered to have a specialized function. This usually necessitates some specialized capabilities, which become highly developed and include specialized information, controls, and autonomous developmental tendencies. Other subsystems tend to become dependent on these specialized subsystems for relief of particular stresses (needs). In this way, subsystems become dependent on one another as well as on the system as a whole.

The whole process of creative synthesis, producing new levels of complex entities, is made possible by the emergence of novel characteristics. These characteristics make possible, or need, bonding (relations), which produces a higher level system. It is also true that large complex systems tend to multiply internal supporting subsystems, specialized subsystems that are developed for some particular internal need (stress) of the systems or some newly perceived external problem. In this respect, social and psychic systems are more adaptable than biological systems. In fact, the production of specialized subsystems that produce a specialized body of information, in turn encouraging the recognition of other new problems, is probably the underlying process increasing complexity in developing societies.

There appears to be a fairly direct relation between the amount of information and the complexity of a social system, and between the amount of information and the rate of development. The more information, the faster the development. Melko's notion of problem perception or problem recognition appears to be more fundamental than Quigley's model, which emphasizes economics and productivity. A model based on information and problem recognition easily includes economics and productivity but can also give full weight to social, religious, philosophical, artistic, and political concerns. The traditional tendency toward identifying some single cause for the evolution of social systems has always been misleading. A simpler and more general model is that a system responds to information with some kind of action. Information input is a stress to which a system responds. All systems continuously encounter various forms of matter and energy. Simple systems perceive only a few of these kinds of matter–energy as information. The more complex the system, and the more specialized the subsystems, the more forms of matter–energy will be perceived as information (stress) requiring some response. With psychic and social systems, the response may well be the creation of another specialized subsystem.

However, a caution must be added here to avoid needless misunderstanding. Complex macrosocial systems (civilizations) do not multiply subsystems and produce specialized subsystems automatically or from some inner predetermination. They respond in these ways to stress. Melko's emphasis on problem perception, or better, problem recognition, reminds us that there may be stresses on systems that are not perceived, not recognized, as stresses, and perhaps not perceived at all by the control subsystems of a system. The energy–matter–information may be beyond the range of the system's receptors. It may be received as meaningless noise or distorted for assimilation into existing maps which interpret it as routine stress or as insignificant, that is, not requiring a response beyond mere recognition. Destructive radiation is a case in point, as are the chemicals with which we are destroying much of our environment and many individuals. There is no necessary conscious awareness due to exposure to destructive radiation or to ingesting lethal drugs. There is, no doubt, awareness at some subsystem levels such as cells or organelles, but the range of receptors that collect information for the central nervous system are not sensitive to that level of energy–matter–information. Another common illustration is a social system that drains the soil of nutrients or lost the topsoil altogether through deforestation, and interprets the information of decreasing productivity as indicating the displeasure of the gods or evil spirits, or as punishment for moral deviance of some sort.

This caution kept in mind, it still appears true that, when perceived as information, stress calls for a response. The response may involve the production of specialized subsystems. If it does, it will also involve the development of greater variety within the system along with specialized bodies of information. This usually means sensitivity to new varieties of energy–matter–information and, consequently, the perception of problems that would not otherwise have been received as information.

Many variables can and do influence the pattern just described. The degree and efficacy of system dominance and control, the degree of common mapping that is maintained in spite of specialization, the amount of energy available to subsystems (e.g., affluence), the degree of plurality among official and unofficial hierarchies, the vitality of the central cultural mythos or religion, or the presence of some degree of threat from the system's external relations (e.g., Pearl Harbor versus Viet Nam, or H. G. Wells' famous radio invasion from Mars) would all make some difference in developmental patterns. These and other conditions would, of course, have to be analyzed in any specific system. The point here is that, when a complex macrosocial system does move through developmental stages, specialization and its attendant increase in the perception of stresses are a central part of the engine.

Central Control System

The two preceding characteristics could lead to fragmentation and disintegration. Central controls are not the only controls, or even the most important under all circumstances. They are essential, however, as they relate to the direction of the use of energy–matter–information in the service of the system as a whole. As Quigley has pointed out, development is related to stored energy–matter–information. Net energy is necessary in any complex system, and requires a central control system with enough power to store it and direct its use. Part of the stored energy–matter–information must be directed toward problem solving, and this tends to support specialization and new dimensions of problem perception. Consequently, there is always a tension for central control systems between problem solution and problem creation. Those in control, at any level, tend to be conservative in relation to "new-fangled ideas." The critical selective process is to differentiate between those solutions that will further the orthodox patterns and those that will introduce antithetical elements. As suggested above, any new solution has the potential for deviant perception resulting from specialized information and tendencies toward increased autonomy.

Some degree of competition always exists among subsystems, subsystem levels, and subsystems and the system. Competition is related to autonomous needs and goals and is for available energy–matter–information. The degree of active competition is relative to the effectiveness of the central control system, which is in turn, relative to the strength of common mapping throughout the system. In early developmental stages, common mapping around a cultural mythos is generally strong and the central control system effective in directing energy–matter–information toward system needs and goals. Unless subjected to interference from other variables, both internal and external development should be strong. When it is not, the researcher has good reason to look carefully for those relations and processes that have blocked the expected patterns.

Changes in the Nature of Bonding

As the processes described as multiplication of subsystems and specialization push the system through successive stages of development, gradual changes take place in the nature of the relations and processes. These changes constitute the system bonds that are both cause and effect of corresponding changes in the information mapping of the system. In early stages, common mapping is very strong, and subsystems tend to define themselves in terms of their participation in the system and its cultural mythos. That is, the psychic systems of individuals and sense of identity of social groups derive their definition, goals, and values from the common system map. The bonds (relations) operate largely through psychic subsystems which we now associate with right brain processes, such as those labeled affective, aesthetic, and holistic. These bonds tend to be personal, involving feelings of loyalty, commitment, and community.

As complexity increases with specialization of information and function, relations (bonds) tend to become more oriented toward left brain processes, that is, the logical, linear, verbal, and technical. There seems to be some correlation between the increasing development of left brain psychic subsystems and the specialization of social subsysems. With this development, bonds (relations) tend to become impersonal, complementary, functional, political, and based on an intellectualized code of controls. Such relations are more easily manipulated and interpreted without consequent emotional reactions such as guilt; hence the concern with the difference between keeping the spirit or the letter of the law. Under these circumstances, central control is weakened and a plurality of interpretations of the common mapping is likely to emerge. A sense of identity then tends to be determined increasingly by the individual and social subsystems, and pressures toward subsystem autonomy increase.

Complex Information Systems

Each of the above characteristics encourages change in the information subsystems at all levels. As already mentioned, information in early stages, is largely common and right brain oriented. That is, the dominant information systems tend to be associational and based on feelings of relatedness. As the system develops, information becomes less common and more differentiated and specialized. Information subsystems tend to become more intellectual, complex, and reflective. All this quite naturally results in the multiplication of psychic subsystems and the fragmentation of information within the system, with the result that communications became more secular (unrelated to the cultural mythos) and critical of the whole system or some aspects of it. This applies to art forms, intellectual activities, and entertainment. One should expect to find a close relation among the processes of secularization, specialization, and increasing complexity of information.

Increasing Subsystem Autonomy

As the above characteristics become more dominant, the central control system and common mapping become weaker as they become more remote in relation to the primary concerns of specialized subsystems. The process appears to be general and gradual through the stages of development of a complex macrosocial system (civilization). However, it may well be more noticeable in transitional periods and later seem to subside under the renewed vigor of a new organizational stage. That is, even if the new arrangement leaves the central control more remote, there can be a greater degree of cooperation toward system needs and goals. The process appears to be a mixture of subsystem specialization and autonomy.

For example, in a fairly pure feudal system, one would expect there to be a stronger participation in the common mapping and much less variety of specialized mapping than in a nation-state system or in an empire. The peasant class in a feudal society would be more directly involved in the understanding of self and world that they share with the nobility and the priesthood. The relations are more direct and so, too, is the dependence. In a nation-state system, social subsystems such as small businesses, guilds, professional associations, and civil servants have more autonomy in terms of the self-interest reflected in their specialized mappings and tend to be more competitive in relation to other social subsystems. In an empire, subsystems tend to become more local and less related to the central and remote-control and information systems.

The increase in subsystem autonomy appears to be related to a number of factors: complexity, variety, specialization, fragmentation of common map, increase in size, and remoteness of central controls. Under these conditions, there is likely to be increasing doubt or lack of concern with the cultural mythos, an increasing number of unorthodox, cultlike forms of salvation offered to alienated or dislocated individuals, and more pluralism in general. Melko's chart of developmental stages make it clear that these tendencies tend to appear in transition stages, especially in going from a state system to an empire. However, the foregoing analysis suggests the continuous development of more variety and more antithetical elements, which are constrained by the power of the emerging order of the next stage, although it is not suggested that one is more fundamental than the other. Increasing subsystem autonomy (differentiation and specialization) eventually forces disintegration or a new level of involvement in a new order. In biological systems, there seem to be alternating periods of growth and stabilization, and in psychic systems, developmental alternations between more and less order and disorder. It would be useful to pursue a study of systems in general to determine whether it is usually the case that periods of fragmenting growth and specialization alternate with periods of increasing integration. Melko's chart clearly fits that pattern.

Stages of Development

There appears to be some common ground emerging among civilizationists. However, in order to avoid needless complications, we shall work from two models, that of Quigley in *The Evolution of Civilizations* and that presented by Melko in pp. 105–107 of *The Nature of Civilizations*.

Quigley's Model

Carrol Quigley describes civilizations as emerging from a stage of mixture in which subsystems with somewhat different mapping and information systems are kept in proximity by some set of relations until bonding (relations) takes place among them. It may be that "kept in proximity" is a matter of communication and transportation. For example, the Greek city-states did not move closer together geographically, but their increased activity among them brought them together in relations. The first external relations among separate or disparate sytems or fragment subsystems initiate the formation of what will become the internal relations of a new social system. In brief outline, Quigley's model lists seven stages:

Mixture

Gestation

Expansion

Conflict

Empire

Decay

Invasion

This model makes good sense in terms of system development. Out of the mixing comes the working out and the ordering of internal relations, boundaries, control systems, increased common mapping and information systems, external relations, and selective criteria for transmission across boundaries. This marks the beginning of a growth process that can only take place when social systems have made the transition from food gathering to food producing. Not all food-producing societies continue to develope. For this to happen, Quigley suggests, they must have an instrument of expansion. This must include (1) an incentive to invest, (2) an accumulation of surplus, and (3) the use of surplus to pay for or utilize new inventions.

If these ingredients are present and functioning, there is gradual development through the period of gestation. In terms of systems thinking, it may be assumed that there will be increasing recognition of stress, which gives rise to specialized solutions that involve new mapping and set up new tensions within the system as a whole, among competing peer subsystems, and up and down the hierarchy. The amount of tension may be negligible or of general significance for the whole system. In any case, gestation is a period of setting the system in order.

Eventually, the recognized stresses will lead beyond problems whose solution is mainly supportive of the forming system to levels of expansion that will carry the system to new levels of development. Throughout gestation and expansion, there are increasing density, increasing subsystem synthesis, increasing specialized information, and increasing control systems. One must assume that new information is either consistent with the general map and an elaboration of it or else not coherent with the general map and temporarily compartmentalized. In the former case, there would be repressed antithetical implications of the new information. In the latter, the new information is largely antithetical and would require accommodating transformation of the general map (cultural mythos, folklore, etc.), but it is compartmentalized. The compartmentalization is probably most often nonrecognition of the conflict, but it remains as part of a gradually growing fragmentation and as a potential source of stress. There are occasional periods of accommodating reintegration through which the cultural mythos (general map) is modified to incorporate some of the new specialized information. This development continues throughout the period of expansion in various urban centers and their peripheral towns, villages, and rural areas.

Expansion may also mean an increase in geographic area, by means of which there evolves a center of development and peripheral areas of development. This leads to a peer-group system without a well-developed central control subsystem. Eventually, the inevitable tension among these peer systems (usually nation-states) is perceived as a stress (problem) that could be relieved through better international relations. The struggle for advantage and power in these relations leads into a stage of conflict, which usually takes the form of imperialist wars. The movement toward empire may be understood as the natural movement toward centralized control systems. In terms of peer-system relations, the stage of conflict involves the determination of which peer system is to become the leading part or central control system. Whether the central control subsystem is political, religious, or of other basis, it is the result of a power struggle and involves the control of the instrument of expansion.

Empire, or the imperial stage, emerges with the resolution of the overt power struggle with the institutionalization of the instrument of expansion. At this stage, the surplus is directed toward nonproductive goals that serve the desires (perceived stresses) of the central control system and the maintenance of the common map.

Quigley says that, at this point, the civilization is overripe. I assume that he means that the mythos or common map has handled all the expansion possible within its interpretive limitations. Modification of the common map is resisted by the inertia and autonomy of the control subsystem until the autonomy of other subsystems increases to produce a weakening of the central control subsystem and the common map. The gradual surfacing of the underlying fragmentation of subsystem maps and information subsystems ushers in the period of decay.

The final stage, of invasion, is dependent on the pressure of a strong peer system in a stage of expansion. This final stage is not, strictly speaking, a stage of development. It might be more instructive to consider what would happen after decay if there were no invasion.

Melko's Model

Melko does not concern himself with the transition from primitive to civilizational social systems, but appears to be satisfied for the present with the observations of Redfield and Coulburn on that subject. He begins with the feudal manor as the basic stage of civilizations. His stages are put most simply:

- Feudal systems
- State systems
- Imperial system
- Decay

However, he carefully describes a transitional stage between each of these, so that the complete model is as follows:

- Crystallized feudal systems
- Feudal systems in transition
- Crystallized state systems
- State systems in transition
- Crystallized imperial systems
- Imperial systems in transition
- Crystallized feudal systems, etc.

The two models are comparable and can be roughly correlated in lists:

Quigley	Melko
Mixture	Feudal systems crystallized
Gestation	Feudal systems in transition
Expansion	State systems crystallized
Conflict	State systems in transition
Empire	Imperial systems crystallized
Decay	Imperial systems in transition
Invasion	Feudal systems crystallized

One advantage of Melko's model is its emphasis on alternating periods of reorganization and stabilization. In this way, it breaks the process down into stages that are more easily analyzed with reference to the causes of fragmentation and reintegration. His model also makes it easy to see that what are called stages are periods of time in which central control and common mapping achieve a higher degree of equilibration than is present during times of transition.

One must assume, from these accounts that the process of change is more or less continuous at various levels of the hierarchy of subsystems. It is not likely to be a process of straightforward, determined, linear development, the mistaken hypothesis of early civilizationists. It is more likely that the processes through which subsystems differentiate and specialize are continuous. The processes through which subsystems are forced to accommodate (conform) to the dominant pattern and sometimes to be disintegrated are also continuous. Each new phase of development for the complex social system as a whole comes about slowly.

The probable pattern is one of many subsystems perceiving problems or stress and correlative solutions that are antithetical to the prevailing pattern of the system. These would normally be managed through institutionalized controls or minor modifications, for example, compartmentalization or partial fragmentation. Antithetical variety is inevitably present in every subsystem, at all levels. In addition, in a developing and changing system, differentiation and specialization continually produce new kinds of antithetical variety, for example, stresses the system does not satisfy or perceptions of other ways of relating. As long as the variety within subsystems is separate and scattered, no unusual or extensive modifications will be required. When the perception of a kind of stress or a source of stress begins to become common mapping among a number of subsystems, however, the conditions for significant fragmentation of the hierarchy of controls and common mapping begin to emerge. If the perception of stress continues to grow and some degree of common mapping begins to unite (bond the subsystems), then serious disintegration may begin and pressure for the new solution will begin to grow. The common mapping of the new arrangement is usually mapped with subjectively different interpretations in each subsystem. The subsystem emerging as the central control in the new system will have sufficient power and influence to persuade or coerce other subsystems to yield to its particular interpretation of the assumed common mapping. When this occurs, the complex social system will appear to have passed through a transition period and arrived at a new stage.

Melko's model traces this movement from relatively high equilibration of a complex social system through a period of relatively high dislocation and partial disintegration. If circumstances are right, a new level of relatively high equilibration may emerge as the next stage of development. If not, further disintegration may follow or the former level of development may temporarily regain sufficient control to keep the system more or less functional. If disintegration continues, Melko suggests, the civilization will return to the level of feudal state systems. However, neither he nor anyone else I am aware of suggests a return to a simple primitive social system. Subsystem mapping does not regress to a level that would reconstruct the mythopoeic participation characteristic of precivilized social systems. It follows that, once left brain functions have attained a certain level of sophistication, social disintegration cannot destroy these functions. Much information may be lost, and new generations educated (informed) only to relatively simple levels, but even they will not return to the level of finding their identity in the mythical unity of the tribe, nor will they be without some of the mapping left over from the disintegrated social system.

Some Applications of General Systems

The purpose of this discussion is not to compare and evaluate these two models. Rather, accepting their broad outlines as basic information, we shall apply a general systems model. Our purpose is to illustrate its usefulness and to show that it makes some characteristics more clear and relatable. At the same time areas will be indicated where more analytical and empirical information is needed.

One great advantage of large social systems is that they provide the best opportunity for close examination of subsystems and sub-subsystems and their processes. It is then easier to see gradual processes taking place, and, the ideas of degrees of existence and degrees of boundary become easier to explain, because one can see the gradual build-up of a network of relations. The point at which one decides that a network of external relations justifies designating the participating systems as a supersystem is a matter of convention or convenience. If the characteristics of a supposedly complete civilization can be listed, various complex macrosocial systems can be classified on a scale representing the gradual development of those characteristics.

The study of complex macrosocial systems should provide a large scale model, the analysis of which should give some indication of what one would expect to find in systems less easily observed, such as psychic systems. Comparing advances in biological information with the analysis of complex social systems provides some clear indications of what one should expect to find in psychic systems, which are probably more complex than the former and less complex than the latter.

A period of mixture in which more than one culture (social map) is involved must be understood as a process in which the antithetical and constrained variety of subsystems from different culture maps find the opportunity to establish new relations. That this can happen indicates the lack, or weakness of, controls in the area of mixing and the presence of stress for which there is no established solution. It also indicates that the different social maps have, or gradually develop, enough common mapping to allow for some degree of common problem recognition and some common mapping of a novel solution. As novel solutions emerge, they will involve new areas of common mapping, which are likely to have aspects antithetical to the original social maps and will thus create new problems. There will necessarily be a selection process through which those who can function best within the new emerging network of relations will not only survive but tend to rise to some level of subsystem control.

If Melko is right, a feudal system is the first stage in the development of a civilization. This is simply to say that social systems appear in which one subsystem, usually a noble family, emerges as one kind of control subsystem, a member (or members) of a religious order (preservers of the cultural mythos or common map) emerges as another kind of control system, and the rest of the subsystems become involved primarily in production (including new little subsystems). This order is not strikingly different from that of a simple primitive social system except in the level of information present in the control systems and in the type of differentiation among the subsystems. The nobility and the clergy are usually somewhat educated, and the quality of participation found in primitive societies is lacking. The relation of the noble family and peasants is more contractual than participatory, for as a new social system gradually emerges from the period of mixture, a significant change takes place. A social system beyond the stage of food gathering has three primary needs for the relief of stress (routine problem solving): integrative internal relations; supportive, selective, and protective external relations; and processes for production and maintenance. The novel characteristic emerging with the new level of complexity is a subsystem that accumulates and manages excess energy–matter–information. The degree and speed of development into more complex stages is directly related to the amount of excess directed toward creating or inventing new processes for the production of energy–matter–information.

However, there is little evidence to suggest that movement in this direction is necessary or innate. Civilizations have developed at somewhat differing rates, some have not developed, and some have been abortive. Indeed, the study of comparative civilization was rejected by historians and has been slow to develop, primarily because Spengler thought he had discovered a necessary and predetermined pattern and a rate for the development of civilizations. Buckley refers to the biological model as morphostatic, or form fulfilling. On the other hand, he sees social systems as form creating; that is, they create their form as they go as a result of adaptive processes. He uses the term "morphogenic" to distinguish those systems that have no innate, pre-established form. Considerable opinion supports the notion that social systems are so adaptive and so various that the search for inclusive comparable patterns is hopeless. However, contemporary civilizationists seem to be pursuing a middle ground that correlates well with a general systems model. Their model, comparable to Piaget's developmental theory of psychic systems, states that psychic and social systems develop in response to conditions (stress) and so are clearly adaptive. They cannot develop in every conceivable manner, but only through an observable series of stages. Supposing that the stages are necessary, the critical issue is the nature of the necessity. The newer developmental model suggests that it lies in the character of the limits and capacities of the subsystems that constitute successive levels of complexity.

Even atoms and molecules have limits with reference to workable arrangements. For example, the outer shell of an atom cannot hold an electron until the appropriate inner shells have been filled, and some deeper inner shells cannot be filled until the appropriate outer shell has been established. Given the limits and capabilities of humans, it is not surprising that families and small social groups should form in rather similar ways. Of course, all such small social systems are not identical; there is greater variety in human arrangements than in atomic arrangements. The question for the study of civilizations is whether there are consistent patterns within the variety. Comparable arrangements in primitive societies can be observed around the world; the same is true of feudal societies, nation-states, and empires. Neither is it necessary or to be expected that each social system would be identical with every other social system in its class. Humans are not atoms, and societies are not molecules. Even biological classes do not include identical members.

From the perspective of general sysems theory, however, it would be a most amazing anomaly if social systems did not exhibit classifiable similarities at different levels of complexity. It is considered useful to classify whales as mammals because this communicates that a particular animal breathes air, reproduces by means of live births, and suckles it young. The classification helps to organize information and provide a perspective. It also sharpens awareness concerning actual similarities and differences. Nearly the same observations can be made about the classification of social systems as primitive or feudal, nation-states, or empires. No classification of systems, from cells on up, can predict the nature or behavior of any member of a particular class. However, what we are learning about systems in general supports the notion that the variations are within limits. Too great a variation generally proves unworkable and leads to disintegration or transformation. There have been anomalies, but, in all systems, they tend to be replaced by the more common and successful kinds of system. When an anomaly is discovered that does persist and is repeated, then a neglected classification has been discovered.

Process of Development

The development of systems from simple to complex requires energy, both for the development and for maintenance. Throughout the universe of systems, the reason for such improbable development is stress; that is, some kind of pressure or problem is relieved or satisfied by investing energy in a more complex system. In such situations, procuring the additional energy creates less stress than attempting to function without it. A lone individual requires less food and shelter than a group. However, the satisfaction of biological drives and the advantages of a group with reference to security and hunting big game encourage the creation and support of social systems. The integration of small groups into simple, primitive social systems appears to be very ancient and very efficient. Movement from that simple highly stable form to more complex levels of system organization comes in response to stress, when, in relation to available information, it is perceived that some more complex arrangement will relieve some discomfort or satisfy some need. Such development requires additional energy for its creation and maintenance, and the use of this additional energy necessitates its production, storage, and distribution. In other words, a controlling subsystem is needed with the power to organize other subsystems for production, hold some products in reserve, and use this reserve to relieve stresses or problems the system may encounter or recognize. This subsystem for the control of production, accumulation of excess, and distribution for relief of stress is essentially what Quigley calls the means of expansion. At the same time, it is consistent with Melko's emphasis on the dialectic between problem recognition and problem creation.

If this control subsystem is accepted as part of the engine that drives the developmental process of complex macrosocial systems, then problem recognition is the fuel. Another part of the engine is the proliferation of specialized subsystems and their specialized information maps. Such specialization is a necessary part of increased complexity in a social system and increases the amount of constrained variety in the system. Every subsystem has potentials for development that must be constrained because they are antithetical to the maintenance of the system, and specialized systems, because of their specialized information maps (perspectives), are likely to have more variety. This generally means that the subsystems will "see things differently" and will tend to recognize problems not recognized by the system as a whole or by other subsystems. If the problems are recognized by the system, or the subsystems are not constrained, further development is likely. Thus, at every level of subsystems, especially specialized subsystems, there is another part of the engine, and another source of fuel, for driving the system along the road to greater complexity. The central control subsystem directing production, storage, and utilization of energy–matter–information is duplicated in some degree within each subsystem as it develops its own needs and goals relative to its own problem perception. Despite the aims and goals of the central control system, the development of specialized subsystems produces some degree of fragmentation. An additional part of the engine derives from stress on the system arising from the threat of disintegration. Maintaining integration is a problem perceived by the system and shared by the various subsystems in varying degrees. It also requires additional specialized subsystems for control, information, and distribution. These subsystems in turn develop specialized information and maps that increase the variety within the system.

These considerations make clear the fundamental tension between system autonomy and subsystem autonomy and the need for constant equilibration at every level in the hierarchies of subsystems. Two different kinds of development take place at the same time: those developmental changes that elaborate the original map of the central control subsystem, and those that do not fit and thus create pressure for modification or transformation. The central control subsystem is not the only source of constraints on subsystems, all of which press toward fragmentation or transformation of the system. On the contrary, the reality of common mapping implies a duality in each subsystem at every level. Common mapping inclines the subsystems toward perceiving their well-being as closely related to participation within the system. Specialized mapping inclines them to perceive their well-being in some rearrangement within the system or in some transformation of the system. There is the potential in every subsystem for mapping the entire system as though the relations were reversed and the system and all its subsystems were to become subsystems. There is a tendency for the particular subsystem to perceive the urgency of its needs as though it were the inclusive system for which all the rest exist and function. This may happen at the level of psychic systems and is clearly true for rival nation-states pressing toward an imperial system.

Varieties of Equilibration

Equilibration of all these stresses can take place in a number of ways. There is, at one extreme, stronger system dominance and resistance to change through increased constraints and a reaffirmation of the primary importance of common mapping (the cultural mythos) and, at the other fragmentation, disintegration, or transformation. Development to a higher level of complexity is an equilibration somewhere between. In social systems, this usually means more inclusive integration with more diversity. The system becomes larger through internal development resulting from increasing proliferation, differentiation, and specialization of subsystems. It also grows larger by annexing neighboring systems.

Within a successful feudal manor, for example, specialization tends to increase with population. There is increasing need for the convenience of craftsmen. If the noble family (the central control system for production) perceives the economic advantage of a more efficient operation, more skilled individuals may be attracted, and the beginnings of a town may form. A town becomes necessary because specialized subsystems need the products (output) of other specialized subsystems. If the group becomes large and dense enough, it will begin to develop the characteristics of a subsystem of subsystems, that is, some regulatory functions will appear as necessary and control subsystems will emerge. Within the controls of the town subsystem, the members of a particular craft may join together to put stress on the town subsystem in order to get a larger share of the available energy–matter–information or they may join together to use a stress to avoid increased taxation. Property owners may do the same. As specialization becomes more developed and successful, there may well be excess beyond what is needed locally and required by the town subsystem and the feudal system. Trade may increase (external relations) and introduce more variety (foreign products), which may create new problems (desires) that can be solved by more production and trade. Tradesmen, business managers, and investors as well as many other specialized subsystems will gradually arise in response to new perceived problems (desires, needs, opportunities, etc.). The process produces a wider and deeper system with increasing numbers and layers of subsystem development. Each subsystem has its own interests and an interest in the whole system. Competition and cooperation require constant equilibration.

Early in such development, the distribution of actual power among competing subsystems becomes one of the most critical matters. It would make a most revealing study to examine carefully the development and arrangements among various control systems as they emerge and compete. Are there characteristic systemic differences that correlate with the strongest control systems in a given social system, for example, the nobility, clergy, capitalists, trade unions, agrarian landowners, and military leaders? Is there a recognizable order in which these systems emerge and rise to power? Is there a recognizable difference in a system's developmental history that correlates with the relative strength of each? Superficially, the difference between the development of the Greek empire and the Egyptian empire suggests that more than geography is involved.

The use of a general systems model allows the investigator to pull back to a more general perspective and suggests a wide variety of things to look for. One should no longer look for a cause, or even a list of causes, but for a confluence of actions and reactions to stresses (recognition of problems), for control systems that emerge to cooperate and compete. One can look for the lines of stress that arise and descend through competing and cooperating hierarchies. Finally, one can look for the gradual development that outgrows one stage of social system organization and presses for a larger and more inclusive arrangement to accommodate a greater amount of differentiation and specialization. Tribes may unite for common defense, just as feudal lords may move toward a nation-state. There may be pressure for integration among towns in order to avoid trade barriers, just as there may be a pressure toward empire for the same reason. Underlying the whole complex pattern, at every level, is the constant problem of equilibrating individual subsystem advantage and the benefits of participation (subsystem autonomy and system autonomy).

Understanding the specifics in relation to a general systems model opens up interesting questions about psychic systems. The theories of Jung and Freud suggest that a psychic system develops competing and cooperating subsystems at many levels. Piaget's developmental theory suggests a pattern not wholly different from the development of a complex macrosocial system. The increase of knowledge since the Renaissance suggests a pattern of self-perpetuating development similar to that of a civilization. In one sense the scientific method is itself a control system for investing stored energy in creative problem recognition and resolution. Pure science, pure mathematics, and game and simulation theories are the acme of problem-creating processes. Science is a method for bringing "ignor-ance" into awareness.[4] Because social systems are so easily investigated, their careful analysis in relation to a general systems model may enlighten the investigation of many other systems that are more difficult to observe.

Residual Maps

Another direction of possible benefit lies in the investigation of the role of residual maps in the processes connected with social system development. Here the term "residual" refers to aspects of social maps (information or meaning subsystems) that are no longer acceptable or tolerated in a succeeding developmental stage. In effect, they become part of the constrained variety of the system and exert some degree of stress on the system or some of its subsystems. How long do such subsystems or fragments of subsystems remain? What are the processes, the family and small-group cultural rituals, for example, that keep them alive beyond the lifetime of the individual psychic systems that knew them directly? What changes are produced by objective information subsystems, such as libraries, recordings, videotapes, and computer banks? Clearly, residual maps within psychic systems sometimes cause problems when circumstances cause them to surface. A comparative study of the effect of residual information maps left over from former stages of development in both psychic and social systems might be very enlightening. In both cases, the effects should have something to do with system inertia (resistance to change), the rate of change in a developmental process, the strength of the new level of integration, and its tendency toward fragmentation. The understanding of any transitional period is enhanced by information concerning the stress between the old and the new and the process for their equilibration.

Degrees of Boundaries and Degrees of Existence

One of the most important aspects of a general systems model is the emphasis it places on degrees of development. Systems do not spring fully developed into existence. They come into existence through the gradual development of relations. As a system gradually emerges, patterns of differentiation and selection develop to distinguish the systems that participate in a particular set

---

[4]The words "ignore" and "ignorance" come from the same latin word, which means "not to know" or "to disregard." We ignore, or disregard, what we do not know. Scientific method makes us aware of what we were ignoring.

of relations from those that do not. If a group of nobles or nations signs a mutual defense or trade agreement that is inclusive and exclusive, they do not thereby form a nation, but they have acquired one aspect of a nation-state system. They may or may not have others but, if the process continues and more relations are formed, they will gradually acquire more and more aspects of a system. For taxonomic purposes, one must decide what degree of system will justify designating it as a nation. However, clear understanding may be better served by indicating degrees of boundary and degrees of existence.

A boundary is a set of relations by which a system is differentiated from other systems. It is always a matter of degree because no system is completely differentiated; that is, no system is completely free of areas that remain somewhat vague through transmissions across boundaries and relations for which there are no boundaries. A boundary is any characteristic of a system that separates internal relations from external relations such as geography, culture (common mapping), contract, formal membership, or blood relations. All relations are bonds and constraints on the participating systems. These systems function as subsystems within the degree of boundary created by the extent of differentiation and selectivity determined by the relations. When a group of nations form an alliance in order to fight a war, this alliance implies a degree of boundary and a degree of system within which the nations become subsystems to a degree. The alliance may some day become an empire but that requires many more binding relations, and hence many more defining boundaries and degrees of existence.

Questions that arise concerning systems at various levels of complexity would best be answered by a description of the kinds of relations, degrees of boundary, and degrees of existence. When two persons become involved in a compelling conversation, they may well exclude everyone else. A thread of system has already developed between them. If they feel a need to see each other again and renew the conversation, they have a relation that already constitutes a constraint. If the conversations continue and satisfy some deeply felt need, more relations will develop between more subsystem levels in their two psychic systems. At some point, they may make a commitment to an exclusive emotional or sexual relationship and may even formalize that relationship with a marriage contract. It is not very useful to try to determine when they constituted a system. They constituted some degree of system from their first conversation. It is more helpful to describe the aspects of their existences that form a system and those that do not. Even after they are married and have many children, their individual existences are still not wholly contained in the familial system.

Behind this line of argument lies the need to shift from a stable, structure-oriented model to a process model. At least since the ancient Greeks, people have been in search of substantive generalizations about things, which Aristotle said either exist or do not exist. Centuries have been devoted to the accurate definition of nouns that were supposed to classify eternal species and even to have a counterpart in one kind of heaven or another. Reality is not like this, but it is hard to break the habit.

World Social System

Are we now in the stages of mixture and gestation of an emerging world social system? Every empire has been an attempt at a world social empire, and all have collapsed sooner or later. A general systems model, with emphasis on degrees of system, provides a grid for the comparison of the kinds of relations, boundaries, and degrees of existence both for past empires and in our present situation. I shall go no further than to suggest that a world system already exists in some degree. In technology, for example, there is a rapidly spreading common mapping. In space exploration and its attendant military capabilities, there is an area of common mapping in which various cultures are eager to participate. In business, there is not only common mapping but powerful interdependence that already encompasses the earth. In mass electronic communication, there is an increasing immediacy to the knowledge one culture has of another. In education, there is a continual flow of subsystems from culture to culture. It is not useful to speculate whether there will be a world social system. One already exists in some degree. It will be more useful to consider in what relations it may continue to integrate or disintegrate and what the consequences of each might be.

Under what circumstances could there be a self-sustaining global social system? In general, the larger systems become, the more variety they must contain within various kinds of hierarchies. Variety is now being manifested in the rise of nationalism around the world. Is this inconsistent with the growth of a world social system? It could be the expression of the disintegrative forces that are always at work in every system. It could be a proliferation of subsystems encouraged by the internal relations of the world social system that already exists in some degree.

Chapter 12

General Systems Theory as Philosophy

In the preceding chapters, a case has been made for the claim that a general systems perspective is inherent in the various specialized fields of knowledge. This perspective widens into a philosophic overview based on the isomorphic relations, processes, and patterns of organization essential to all entities, in every field of study. There is a rapidly growing awareness that the "entities" with which human knowledge is concerned are relations of organized processes. Gradually, the notions of *substance* and *thing* are being replaced by *organization* and *process*. With the description of the classes of relations, processes, and patterns of organization characteristic of all forms of existence, it becomes evident that novel relations, processes, and patterns of organization emerge at each level of complexity. These emerging characteristics seem to be universal for all entities of the level at which they emerge and for all more complex levels synthesized from them

A philosophy of process and a hierarchical views of the universe are not the creations of general systems theory. In 1929, Alfred North Whitehead published his major work, *Process and Reality*,[1] which established the patterns for thinking of the universe as events rather than things. A similar kind of thinking has gradually emerged under different titles and in different contexts. Werner Heisenberg concludes in *Physics and Philosophy*[2] that, in place of things, there are connections and that these connections are in process throughout the universe. An interesting variation of the same general perspective appears in Teilhard de Chardin's *The Phenomenon of Man* and Paul Tillich's *Systematic Theology*.[3] This kind of thinking has been "in the

---

[1] Whitehead, Alfred North, 1929. *Process and Reality*. New York: The Social Science Book Store.
[2] Heisenberg, Werner, 1958. *Physics and Philosophy*. New York: Harper and Brothers.
[3] Chardin, Pierre Teilhard de, 1959. *The Phenomenon of Man*. New York: Harper and Brothers.
Tillich, Paul, 1951. *Systematic Theology*, 2 Vols. Chicago: University of Chicago Press.

air." It has appeared in introductory chapters in textbooks for science courses and in such special publications as Clifford Grobstein's *The Strategy of Life*.[4] This same basic perspective has been developing through the study of particular systems and, since about 1940, as a theory of systems in general. Systems science and systems analysis have grown very rapidly through various fields of engineering and technology. General systems theory is based on the hypothesis that all systems (man-made, natural, or symbolic) have some characteristics in common and that these common characteristics serve as a description of the nature of the universe or the nature of existence. These characteristics are the foundation for a new model for perceiving the unity of the universe. It provides a new map within which humanity can locate it and a new context within which to make important choices.

Mythopoeic World Views

As far as we know, individuals have always had a holistic view of themselves and their world, except perhaps in periods of major cultural transformation with their attendant psychosocial dislocations. The earliest forms of these total views we call mythologies because they were largely narrative, imaginative, and affective. They did not depend on careful, logical definition of terms, reasoning processes tested by a system of logic or on existential statements verified by a rigorous methodology. They were, nonetheless, synthetic perspectives on the nature of man and his universe founded on the best information available at the time. That this information was anthropomorphic, allegorical, and superstitious does not alter the fact that some of the forces at work in the human mind, as in all systems, were—and are—integrative. Whatever the foundations of our interpretation of experience and whatever our store of information, we will, if possible, organize it into an overview or a total feeling about ourselves. The human brain, probably the right cerebral hemisphere, has a holistic function as part of its organizing capabilities.

Rational World Views

The great discovery formalized by the Greek philosophers was the power of accurate definition of words and their relations: logic, grammar, and syntax. Socrates sought the answer to some of the relativistic opportunists of his time in the clear definition of the general terms he thought would provide a clear understanding of the nature of a stable reality. Plato believed that existential generalizations could be as clearly defined as mathematical generalizations and that, taken together, they would constitute a formal, stable, and eternal reality. For Plato, all forms of existence are changing, temporary copies of

---

[4]Grobstein, Clifford, 1965. *The Strategy of Life*. San Francisco: Freeman.

these formal perfections. Although he never succeeded in setting down absolute definitions for goodness, beauty, and truth, not to mention men and mud, he did set the stage for Aristotle to formalize the laws of logic. With logic and Pythagorian mathematics, the Western world was on its way to a new foundation for its understanding of humans and their universe, in other words, for its philosophy. The underlying assumption was that generalizations (universals) that are clearly and distinctly understood in accordance with logical consistency are necessarily true. In other words, the test of the truth of a statement became whether the enlightened human mind could understand it clearly and distinctly and with logical accuracy. One might suggest, in the light of recent discoveries, that the center of understanding shifted from the right hemisphere of the human neocortex to the left, from affective, allegorical, and narrative information patterns to those that are linear, logical, and rational. However, the rationalists and idealists did exactly what the ancient mythmakers had done before them. They created information systems that mapped holistic perspectives of the nature of humans and their world. They used the latest and most accurate information available to create an overview within which they could evaluate their lives, make choices, and set goals.

The shift was gradual, and it would be instructive to analyze the varying degrees of intellectual and affective elements in the succession of philosophies that have appeared from ancient Greece to the present in the dramatic struggle of humans to maintain their sense of participation and belonging through stages of cognitive development. Hindu philosophy has analyzed four ways in which one can overcome one's alienation or sense of dislocation: cognition, emotion, activity, and introspective discipline and detachment. Western philosophy has been characterized, for the most part, by a process of equilibration between the polar extremes of rationalism and mysticism as humanity attempted to accommodate new information while maintaining some sort of overview. In this connection, it is interesting to compare the intellectualism of Aristotle with Epicureanism, Stoicism, Skepticism, and the mysticism of Plotinus. Both polar extremes are clearly evident in varieties of Christian philosophy through the Middle Ages. Following the Renaissance, continental Rationalists became convinced that pure reason could solve all the problems of humanity and the universe.

Medieval Christian scholars believed that sound, rational universals were akin to God's thoughts about the universe. Descartes is said to have closed himself in a great stove, or heated room, until he had thought out the true philosophy of humans and the world. Hegel wrote an elaborate philosophy of history without concern for analyzing actual historical patterns. None of this is more surprising than that a computer will follow the relational principles it is fed and organize accordingly the information provided.

People have always depended on experience for information, and their interpretation of that experience led to many highly imaginative and emotionally distorted interpretations of themselves and the universe. At the same time, however, a new stage in human cognitive development was emerging. This new stage began with the recognition that human experience was often deceiving and that rational and emotional processes in the human psyche were seductive and biased. This recognition gave rise to the scientific method for empirical verification and philosophical empiricism. Philosophers began to analyze human experience in relation to the universe much as scientists developed increasingly careful methods for testing and verifying the relation of ideas (generalizations) to actual entities and processes. Human reason was thus gradually removed from the throne and became a function for processing information. One can borrow from the wisdom of those who work with computers and accurately say of human reason, "garbage in, garbage out." Mathematics came gradually to be recognized as having no necessary connection with the entities and processes of the universe. It is now accepted as commonplace that the human mind can create many different symbol systems with various sets of relations and play endless games, some or none of which may have existential applications.

Humanity's love affair with its capacity for accurate, logical, and linear thought processes lasted about 2000 years before interest shifted to empirical verification. It took that much time for people to weary of spinning out varieties of philosophies, all of which served some practical need of a particular time or person, but none of which could be shown to be true or false.

The Fragmentation of Knowledge

The foundation of information gradually shifted from "clear and distinct ideas" to degrees of empirical verification. Over the next 400 years, available and reliable information has increased at an incredible rate. It has become necessary to concentrate research on smaller and smaller areas of investigation. Ideas are no longer considered superior forms of reality. They are elements of one among many symbol systems. Reality is the universe, and we are struggling to learn its secrets through accurate observation and experimentation. However, the time needed to acquire the necessary expertise for functioning on the current frontiers of human knowledge is long relative to the average human lifespan. We continue to learn more and more about less and less. We are all aware of the frustration of trying to keep up with advances in one small field of research and have come to accept the hopelessness of concerning ourselves with several.

One of the consequences of this growth of available, accurate information is that we cannot make much use of it for the understanding and conduct of life. Individuals have a variety of philosophies and religions, or fragments thereof, by which they try to make sense out of their lives and times. I suspect most get along from day to day using programmed or conditioned patterns we usually refer to as common sense or good judgment. The culture, or the civilization, has no philosophy that functions as a synthesis of the best available knowledge, and many suggest that such a synthetic overview is neither possible nor necessary.

Another consequence of the rapid growth of knowledge is the fragmentation of specialized vocabularies, academic departments, and education in general. A recent report from a Carnegie Commission referred to "general education" as a disaster area. One can easily agree with this evaluation if disaster is taken to mean "dis-integrated." There is nothing disastrous about the courses taught in our colleges except that they are unrelated. However, the solutions proposed are so superficial as to miss the essence of the problem altogether.

There is growing concern for interdisciplinary education, and many fine scholars and teachers are working to make it a reality. Much of what they achieve is worthwhile for the students, but they have yet to reach the root of the problem. One teacher, or a team of teachers, can only squeeze so much content into a set amount of time. The needed synthesis cannot be achieved by setting more and more areas of specialized knowledge side by side under some thematic or periodic umbrella. I hope to make it clear that systems thinking can answer this problem for our time.

Civilizational Disintegration

A third consequence of the increasing supply and fragmentation of information is the breakdown of the "cultural myth." Those who work in the area of comparative civilizations agree that there is a stage in civilizational development in which the driving force gives way or runs out and the civilization begins to stabilize relative to growth and disintegration. In this period, the religion or set of interpretive convictions with which the civilization began disintegrates and is replaced by cultlike religions, secular philosophies, skepticism, psychosocial dislocation, and a high degree of pluralism in organization, values, and goals. These conditions hold all around us and, to an increasing degree, around the world.

However, not every civilization that reaches this sort of maturity and begins to disintegrate actually falls apart. The apparent disintegration, accompanied by the above-mentioned symptoms, is often part of a process of transformation rather than destruction. Such is the case when a new source of motivation emerges in the culture. This could be a new religion, philosophy, or technological advance, a strong external threat, or anything else that precipitates a new sense of identity and motivation toward common goals. General systems theory, understood as a philosophy of world and life, could well lay the foundation for such a revitalization in our time. It is becoming abundantly clear that systems thinking will emerge as the operational philosophic context to overcome the isolation of specialists and to resolve the problem of fragmented compartmentalized knowledge that is defeating the traditional purpose of undergraduate higher education.

The Function of Generalizations

The human mind cannot have a separate symbol for each separate entity it experiences. Nouns exist in our symbol systems so that we can process information concerning chairs—or trees, or cars—as a class. The same is true of processes or actions: We have verbs and prepositions to generalize relational connections. Taxonomy is a necessary part of any growing science or body of knowledge. Reliable generalizations are the result of careful work and much testing. We have learned to be skeptical of the easy generalizations that are always in abundant supply. They are the stock-in-trade of advertisers, promoters, politicians, and those who would sell us passage into one utopia or another.

A genuine synthesis of our fragmented specializations can come about only through the evolution of reliable generalizations that reach across the barriers of specialized vocabularies, methodologies, and models to reveal the unity that underlies all reliable knowledge. Such generalizations, cannot carry the detail of specialized knowledge with them but that is not their purpose. The general term "chair" does not give much detailed knowledge about any particular chair, but it does locate certain objects within a general body of organized information. With this orientation, we can more easily pursue more detailed information about subclasses of chairs or a particular chair. We can also make more intelligent decisions about these objects, their relation to other objects, and their relation to our lives and purposes. Further, upon encountering one of these objects, we have some hooks on which to hang any new information we may acquire about it.

All generalizations contain hidden dangers. The major war that science has waged, and continues to wage, even within its own house, is to break the stranglehold of dogmatic generalizations that are precious parts of accepted interpretive models. The only solution is that practiced by every specialization within its own field: There must be a continual dialectical movement between generalizations as working hypotheses and analytical verification of their accuracy and usefulness at the growing edge of detailed research.

Generalizations revealing the unity of all reliable knowledge must derive from careful investigation of specialized bodies of knowledge. The purpose of such an investigation is to identify those isomorphic characteristics about which operational and reliable generalizations can be made. To the extent that generalizations can be identified that are universally applicable, one can begin to define the conditions of existence and the essential nature of the universe as revealed by the contemporary state of knowledge.

This may sound like an outrageous suggestion, but from the perspective of general systems theory it is a normal process in human history and inevitable short of a major catastrophe. The synthetic process is well on its way. Universal generalizations are emerging that will need testing and refining. The main barrier is the compartmentalized mind set of those who are needed for the dialectical process of hypothesizing and testing. Eventually, as these generalizations are organized, they will provide a new model of humanity and the world and a new context within which we can struggle with the basic human problems.

Relativism

Prior to the rapid growth of science, it was not uncommon for philosophers and theologians to regard some generalizations as absolutes. In terms of individual psychology, many generalizations still function as absolutes. However, the growing demand for empirical verification, the increasing sophistication of the natural and social sciences, the advances in mathematics, and the even wider application of scientific rigor to new areas of human experience have gradually toppled one absolute after another. The emergence of a heliocentric interpretation was followed by Einstein's theory of relativity; the emergence of an evolutionary view of man, human society, and the universe; and the application of historical and textual criticism to sacred, historical, and literary documents, giving rise to something called "relativism." In the popular mind this became a kind of informal libertarian philosophy of life or indicated that no philosophy of world and life was possible or needed. The conviction of dedicated researchers to follow the evidence wherever it led was translated into the individual's right, even obligation, to act on the basis of expedience and popularity, or to search for the true emancipated self.

Serious theories that point to the relative character of everything in the universe were and are concerned with analyzing the laws and principles that govern those "relativities." They suggest to many, however, the more general cultural relativism implicit in statements like "God is Dead." If any absolute remains, it is a notion of individual freedom, which in practice often enforces the attempt to avoid any kind of constraining relations. In the void that follows the breaking of relations, it is not surprising that paths to individual salvation, escape into drugs, communes, and cults abound. Nor is it surprising that contemporary literature reveals an overriding concern with anxiety, meaninglessness, and despair. The relative quantities of the universe are ordered, lawful, and interrelated, the laws revealed in scientific investigation must at any given time function as absolutes, but the laws are *relational* absolutes. Even the law of gravity, a fairly convincing absolute, is relative in the sense that its application depends on the masses of the related bodies. The god who died is the god of a static stable universe in which there were substantive absolutes with supposedly predetermined applications to humans and their world. Emerging from the temporary fragmentation and partial disintegration of a major transformation is a view of the universe that is steady rather than stable and universals or generalizations that are relational rather than substantive with applications that are relative but lawful and largely determined.

Relational Universals

Terminology is always a problem when one is trying to break new ground. The term "relational universal" will be used here to refer to generalizations that can be made about everything that exists. I shall call these universals "relational" because they concern the relations among entities. A perspective that regards entities such as tables as "things in themselves" will probably demand what are called *substantive* universals or generalizations. A perspective that regards entities as sets of relations will accept *relational* universals or generalizations. Indeed, one soon discovers that in specific application the case is similar to what Orwell described in *Animal Farm*: Some universals are more universal than others.

There are some generalizations true of any and all systems—of everything in the universe, although I hesitate to call them absolutes. These relational universals appear to be the necessary conditions—the definition—of existence: To be is to be related. In other words, that concerning which no relation is known cannot be known to exist. If all possible kinds of relations were known, all possible universal truths about the universe would be known, but that will never be the case.

Mathematics: The Pure Science of Possible Relations

Mathematics is the science and study of pure relations. Its content is a network of relations based on the definition of symbols and fixed rules for manipulating them. A given quadratic equation indicates many sets of relations. When specific numbers are plugged in, a specific set of relations results that may be interpreted as representing an existential set of relations. Taken together, such equations provide the relational universals for existential processes.

General Systems Theory: An Existential Science of Possible Relations

General systems thinking is one step less abstract than mathematics; that is, it is directly related to the analysis of existential systems. In this sense, it is closer to applied mathematics and, in many areas of system analysis, the two appear to become one. However, systems thinking must of necessity use a symbolic information system that is more inclusive than mathematics or computer language, the ordinary languages in which most of our information, thought processes, and communications are symbolized. General systems

General Systems Theory: An Existential Science of Possible Relations 219

theory is the quest for relational universals that are true for systems in general, relational universals that emerge at new levels of complexity, and a model of the whole of existence as the interaction of entropic and negentropic processes.

General systems theory as a philosophy of world and life makes certain assumptions and draws certain inferences from the present state of knowledge.

1. It is assumed that the universe is ontologically continuous and coherent, that is, a unified whole of interrelated parts.
2. It is further assumed that accurate knowledge of the universe, despite the diversity of specializations and vocabularies, has an underlying epistomological unity that models the nature of the unity of the universe.
3. It is inferred from our present state of knowledge that the universe is constituted entirely of organized entities that contain organized parts and participate in the organization of larger wholes, entities ranging from pure electromagnetic energy to the universe as a whole. Because the term "organized" means standing together in sets of limiting, constraining, and controlling relations, these entities are properly called systems. Hence, the universe is inferred to be a system of interrelated systems.
4. Current knowledge recognizes that the universe is a hierarchy of systems; that is, simple systems are synthesized into more complex systems from subatomic particles to civilizations.
5. All systems, or forms of organization, have some characteristics in common, and it is assumed that statements concerning these characteristics are universally applicable generalizations.
6. All levels of systems have novel characteristics that apply universally up the hierarchy to more complex levels but not down to more simple levels. Universals have a floor below which they do not apply but they apparently have no ceiling.
7. These characteristics are always relations of one kind or another. Hence, they should be called relational universals, to distinguish them from an older generation of universals that named existing entities and tried to constitute their taxonomy. The latter may be distinguished by the name "substantive universals."
8. Because the focus of any investigation is limited in relation to the levels, complexity, and diversity of actual systems, it is necessary to have general terms that can be applied to any selected level on the hierarchy of systems, and the terms "system," "subsystem," and "supersystem" should not be associated with definitions peculiar to one level of system or another. They must be given definitions general enough to be applicable to any level of system. It is only in this way that they can be useful in representing the underlying unity of the universe and the consequent unity of knowledge.
9. It is possible to identify relational universals that are applicable to all systems at all levels of existence.
    a. Every system has a set (or sets) of internal relations. These relations

are constraints on the potential variety of its subsystems, on the development of the system, and on the variety of external relations. They also determine some of the system's boundaries.
b. Every system has a set (or sets) of external relations. These relations are mutual constraints within supersystems, among peer systems, and within the universal supersytem that limit or determine the behavior and development of the system. These relations, in conjunction with internal relations, determine the system's boundaries.
c. Every system has a set of boundaries that indicates some degree of differentiation between what is included and what is excluded from the system.
d. Every system is selective among possible relations and with reference to transmission across its boundaries.
e. All selectivity is based on information within the system (mapping), even if the information is nothing more than the selective and discriminating nature of its relations and boundaries.
f. Every system has some degree of inertia; that is, every system resists some kinds of change to some degree.
g. Every system is constituted of processes of equilibration, where the word "process" indicates relations that are, or are part of, sequences of change. Equilibration processes, the processes that balance and constrain one another, are control subsystems or information systems determining formal or existential patterns. Functions are processes perceived in relation to the needs of a system. Equilibration processes may be routine modifications of some aspects of the system. Modifications may be some form of development toward more complexity or disintegration toward less complexity. Disintegration of the system into its component subsystems is a form of equilibration in a larger system context.
h. Every system has some form of control that tends to maintain its integrity. These controls may be any equilibration processes of the forms just mentioned.
i. Every system is subject to stress, used here in the general sense of any form of energy–matter–information that relates to a system so as to produce activity.
j. Everything that exists, whether formal, existential, or psychological, is an organized system of energy, matter, and information.
k. Every system is a set of relations involving energy, matter, and information.
l. If energy is understood as anything that tends to produce motion or work, or to activate processes, then energy, matter, and information are indistinguishable in some of their relations.
m. Every system contains internal stresses that are equilibrated through internal processes. Some of these stresses can be represented as universal polarities:

(1) System dominance and subsystem autonomy.
(2) Selective sytem controls and subsystem variety. No system or subsystem is completely distributed with reference to its potential for relations or development. Within any inclusive system or combination of inclusive systems, there are always antithetical elements. Every subsystem has a tendency to develop in relation to its autonomous needs. There is always constrained variety.
(3) Processes of integration and disintegration.
(4) Processes of synthesis and emergent novelty (e.g., specialization).
(5) Processes of development and processes of conservation (inertia).

n. Every system is subject to stresses through external relations. Since every system is related to other systems and to the entire universe as system through its internal relations, its external relations are the internal relations of some more inclusive system or systems and participate in the polarities listed above.

o. External relations may be relations within an hierarchy, among multiple hierarchies, or among peer systems. They may also be occasional encounters among otherwise remotely related systems.

p. The universe consists of processes of synthesizing systems of systems and disintegrating systems of systems. It will continue in its present form so long as one set of processes does not eliminate the other. (Both the Big Bang and Steady-State astronomical theories are consistent with this polarity.)

These assumptions and statements constitute a model of the nature of the universe and the conditions for any kind of existence. They are derived by beginning at the highest level of generalization, and provide a framework for research into the areas of specialized knowledge for verification and improvement of general applicability. They have heuristic value for problem solving, because they indicate relations that should be researched, and serve as a corrective for models too narrowly conceived. They provide an alternative to older models, which have created a number of needless problems.

This model appears to be applicable to all systems, whether formal (e.g., mathematics and language), existential (e.g., "real world" systems), or affective (e.g., aesthetic, emotional, and imaginative systems). In this sense, the problem of ontology (what is subjective as opposed to objective) is replaced by statements concerning relations (successful or unsuccessful) among different kinds of systems, for example, formal and existential, or psychological and existential. Such a relational ontology is concerned not with distinctions between the "real world" and formal abstract systems, but with describing relations within different kinds of systems and degrees and kinds of relations among different kinds of systems. From this perspective, general systems research or theory is simply some number of steps more general than specialized areas of knowledge.

Reductionist questions are replaced by a description of the external relations

of a system as distinguished from the external relations of any c[...]
systems. That is, the sum of the parts has external relations that [...]
cannot have except as that sum.

This model provides a context within which fragmented knowl[...] education can be reunited, in the humanities, arts, sciences, an[...] disciplines. Finally, it provides a framework within which huma[...] locate itself and the world and within which the ancient human conc[...] be redefined in more manageable terms.

Some Implications

The universe consists of constantly repeated processes of integration and disintegration. It is energy; particles, atoms, molecules, and more complex interrelated systems; and processes through which energy and systems are transformed into one another and back again. One direction is negentropic, as from particle clouds to stars. The other is entropic, from complex matter systems to radiant energy. There is constant integration and disintegration as the universal processes create stresses that are equilibrated. A new form of sytemic or disintegrative equilibration will, in turn, create new stresses within the universe that, again, will produce new processes of equilibration. This has been the case for what appears to be some 15 billion years, since the Big Bang. If the expansion of the universe means that entropy is increasing on the whole faster than the equilibrating negentropic processes, then there may be a reversal, as some astronomic speculations suggest, when the universe will draw together again preparing for another bang. In the meantime, the universe appears to be a two-way process equilibrating between integration and disintegration.

If we narrow our field to the earth–sun system, we find an interesting variation. On earth, these universal, processes, fed by the abundant energy of the sun, have continued in two directions, but with increasing amounts of energy used and stored in the creation and maintenance of very complex systems, from nucleotides to civilizations. The evolution has been slow, but emergent novelty has been one of the main features since the processes have shown definite stages rather than proceding in the direction of negentropy. There have been 7 or 8 million years of novelty and increasing complexity, 2 million years of human life, and perhaps 8000 years of civilization.

Hierarchy of Complexity and Novelty

The evolution of systems on the earth has been a process of synthesizing systems with the potential for novel external relations (bonding) that were finally drawn into more complex systems. At each new level, those new complex systems that were successful (survived) possessed the potential for new kinds of external relations. On the cooling earth, subatomic particles of a system as distinguished from the external relations of any of its subsystems. That is, the sum of the parts has external relations that the parts cannot have except as that sum.

This model provides a context within which fragmented knowledge and education can be reunited, in the humanities, arts, sciences, and formal disciplines. Finally, it provides a framework within which humanity can locate itself and the world and within which the ancient human concerns can be redefined in more manageable terms.

Some Implications

The universe consists of constantly repeated processes of integration and disintegration. It is energy; particles, atoms, molecules, and more complex interrelated systems; and processes through which energy and systems are transformed into one another and back again. One direction is negentropic, as from particle clouds to stars. The other is entropic, from complex matter systems to radiant energy. There is constant integration and disintegration as the universal processes create stresses that are equilibrated. A new form of sytemic or disintegrative equilibration will, in turn, create new stresses within the universe that, again, will produce new processes of equilibration. This has been the case for what appears to be some 15 billion years, since the Big Bang. If the expansion of the universe means that entropy is increasing on the whole faster than the equilibrating negentropic processes, then there may be a reversal, as some astronomic speculations suggest, when the universe will draw together again preparing for another bang. In the meantime, the universe appears to be a two-way process equilibrating between integration and disintegration.

If we narrow our field to the earth–sun system, we find an interesting variation. On earth, these universal, processes, fed by the abundant energy of the sun, have continued in two directions, but with increasing amounts of energy used and stored in the creation and maintenance of very complex systems, from nucleotides to civilizations. The evolution has been slow, but emergent novelty has been one of the main features since the processes have shown definite stages rather than proceding in the direction of negentropy. There have been 7 or 8 million years of novelty and increasing complexity, 2 million years of human life, and perhaps 8000 years of civilization.

Hierarchy of Complexity and Novelty

The evolution of systems on the earth has been a process of synthesizing systems with the potential for novel external relations (bonding) that were finally drawn into more complex systems. At each new level, those new complex systems that were successful (survived) possessed the potential for new kinds of external relations. On the cooling earth, subatomic particles

(1) System dominance and subsystem autonomy.
(2) Selective sytem controls and subsystem variety. No system or subsystem is completely distributed with reference to its potential for relations or development. Within any inclusive system or combination of inclusive systems, there are always antithetical elements. Every subsystem has a tendency to develop in relation to its autonomous needs. There is always constrained variety.
(3) Processes of integration and disintegration.
(4) Processes of synthesis and emergent novelty (e.g., specialization).
(5) Processes of development and processes of conservation (inertia).

n. Every system is subject to stresses through external relations. Since every system is related to other systems and to the entire universe as system through its internal relations, its external relations are the internal relations of some more inclusive system or systems and participate in the polarities listed above.

o. External relations may be relations within an hierarchy, among multiple hierarchies, or among peer systems. They may also be occasional encounters among otherwise remotely related systems.

p. The universe consists of processes of synthesizing systems of systems and disintegrating systems of systems. It will continue in its present form so long as one set of processes does not eliminate the other. (Both the Big Bang and Steady-State astronomical theories are consistent with this polarity.)

These assumptions and statements constitute a model of the nature of the universe and the conditions for any kind of existence. They are derived by beginning at the highest level of generalization, and provide a framework for research into the areas of specialized knowledge for verification and improvement of general applicability. They have heuristic value for problem solving, because they indicate relations that should be researched, and serve as a corrective for models too narrowly conceived. They provide an alternative to older models, which have created a number of needless problems.

This model appears to be applicable to all systems, whether formal (e.g., mathematics and language), existential (e.g., "real world" systems), or affective (e.g., aesthetic, emotional, and imaginative systems). In this sense, the problem of ontology (what is subjective as opposed to objective) is replaced by statements concerning relations (successful or unsuccessful) among different kinds of systems, for example, formal and existential, or psychological and existential. Such a relational ontology is concerned not with distinctions between the "real world" and formal abstract systems, but with describing relations within different kinds of systems and degrees and kinds of relations among different kinds of systems. From this perspective, general systems research or theory is simply some number of steps more general than specialized areas of knowledge.

Reductionist questions are replaced by a description of the external relations were bonded to nuclei and became atoms through the equilibrating processes of kinetic energy and nuclear attraction, and atoms formed electrostatic and covalent bonds to form molecules. The process continued, with the aid of abundant energy from the sun, forming macromolecules, cells, organisms, and so on, up to people, social systems, and civilizations. Through the interaction of human psychic and social systems, the universe continues to produce novel systems of matter, machines, and communications. The whole system of systems on the earth is a hierarchy in which the more complex systems are composed, level by level, of more simple systems. The process cannot move from subatomic particles to molecules without the novel external relations of atoms. Cells cannot form without nucleic acids, just as nucleic acids cannot form without molecules of sugar, nucleotides, and phosphate groups. Similarly, the human psychic system can reach the capacity for abstract reasoning only by building from simple systems of sensorimotor skills through layers of information organization to systems of abstract symbols. Social and psychic systems have evolved slowly through gradual accommodations between individual and social subsystems and their inclusive social systems.

The process has been one of synthesis, stress, equilibration, disintegration, and synthesis once again. The process depends on the proximity of systems (with the proper potential for external relations) over a long enough time and under the right conditions for bonds to form and synthesize a new entity. The same is true for complex molecules in the primordial ocean, chemicals in a test tube over a Bunsen burner, humans from various cultures gathering in a fertile river valley, or bits of information in my psychic system trying to synthesize the communication contained in this book.

The Open-Ended Universe

We do not know what comparable negentropic processes may be going on in other parts of the universe, but from our understanding of the process on earth we may suppose that the universe is open-ended in the direction of evolving complexity. The more complex the system, the larger the requirements for energy, but the abundance of energy being radiated from all the suns of all the galaxies suggests that the limits have not been tested yet. Further, the pattern of emergent novel characteristics suggests that the universe is unpredictable with reference to a new level of complex systems. It is not likely that an exhaustive study of atoms of hydrogen and oxygen could have predicted the characteristics of water without knowledge of some comparable synthetic system on the same level. Even then, the prediction would have been only a calculated guess. In short, a general systems view reverses the old philosophic self-evident truth that the cause must be greater than the effect. On earth, the effects are continuously greater than the cause, as long as great is equated with negentropic complexity.

All these observations taken together as a model or a philosophy of world and life have significant implications for an understanding of ourselves and our universe. A few important problems may be selected as illustrations.

Our Place in the Universe

It has been rather common throughout the history of folklore and philosophy for people to assume that they and their social system were at the center of the universe symbolically, ideologically, and geographically. This is not surprising when one considers that our conscious life evolves as an observing center storing information about the environment that surrounds it, its own inner reactions, and the best way to relate one to the other. Piaget gives a convincing description of the process of subsystem building by which children dissociate what is self from the general field of experience. In systems terminology, the emerging awareness of the psychic system gradually distinguishes developing subsystems of internal relations and correlates them with developing information maps of external systems. The pattern of inner and outer, like that of figure and ground, makes it appear reasonable that the individual psychic system would emerge with an orientation of a self and social system surrounded by a universe of entities. These suggestions are supported by mythologies and theologies from around the world explaining how a god or group of gods created their country for some special reason. Invariably, these gods are very concerned about humanity and everything it does. They manage the forces of nature directly in response to the way humans relate to them and their desires. The so-called great religions of the world, where one finds a more intellectualized description of god, gods, or an ultimate ground of being and a more sophisticated system of ritual and ethical behavior, nonetheless still consider humans the center and focus of cosmic and metaphysical happenings. We shall later consider some fascinating exceptions to this general pattern.

As a result of the gradual shift in our understanding of ourselves and our universe, humanity now appears to be a tiny living system inhabiting a speck of cosmic dust that orbits a mediocre star about two-thirds of the way out from the center on one arm of one galaxy somewhere in an expanding universe of numberless galaxies. From this perspective, we are not the center of the universe in any terms, and our existence is of no measurable consequence to cosmic processes. This shift in models continues to produce a variety of reactions in human psychic systems. One could well conclude that humanity is of no consequence in the universe.

However, a different perspective can be derived from the same model if one looks more closely at the human situation. On Earth it appears that the evolution of the hierarchy of systems described above has been the result of interacting processes for which there was no ordered control sytem, processes we commonly call trial and error. They have produced many systems that could not equilibrate internal processes, stress from external processes, or both. However, out of the great variety of systems produced over an abundance of time, a hierarchy of successful systems has emerged. At the top of this hierarchy, an adaptive system has emerged with a unique reflexive feedback system that allows it to function at several levels of awareness and abstract function. Consciousness, as it occurs in the human psychic system, is a novel emergent on the earth. As far as we know, the universe has become conscious of itself for the first time through human psychic systems. (If there are other conscious intelligences on other planets, the statement must simply be expanded to include all such intelligent psychic systems.)

From this perspective, the significance of our role changes. The human psychic system is a synthesizer of new systems and has gradually altered the equilibration processes that evolved in the earth–sun system. General systems theory focuses upon our place as a system within a system of systems. It emphasizes that every adjustment in one system puts some stress on other systems, which must then equilibrate or disintegrate. Humanity has taken over part of the control subsystems of the earth–sun system. We preserve life systems that would not survive otherwise and destroy many living systems that are part of the total equilibration process. We are only now becoming aware of some painful problems. Whoever decides to preserve life is inevitably confronted with the question of when to end it. Things that were formerly left comfortably to God or nature are now clearly a human responsibility. We have become a significant part of the control system, and we are not very comfortable with the role. In cosmic perspective, humanity may be minuscule, but it has become a critical part of the control system of a very complex and highly volatile system.

In addition to the problems of equilibration, systems theory suggests that the universe is open ended in the direction of synthetic novelty. This does not mean that all things are possible; it does imply that there are always other possibilities. There is no cosmic or genetic chart for the proper programming of a human psychic system or for the proper structure and control of social systems. We have open before us a variety of possible futures, and groups of people are now developing sophisticated games for symbolically testing models and variables. I am sure that the Psalmist was not thinking of systems, but he must have had a similar perspective when he wrote, "Yet thou hast made him a little less than God."

Some Problems

General systems theory puts the problem of moral choice and ethical systems in a new light. The various ethical codes by which humans have and do live are attempts to formulate good subsystem relations, that is, to objectify or formalize operational guidelines for the control of the internal relations of a social system. A central problem is that these control systems do not come ready-made, we are always aware of a variety of possibilities. Systems theory cannot provide answers or ideal control systems, but it can direct attention to the analysis of several levels of interaction in terms of which possible consequences can be projected. The unavoidable polarity of subsystem autonomy and system dominance is a reminder that not all variety is possible within any given system. Some possibilities must be constrained. Further, every decision produces some stress on other systems up and down the hierarchies of super- and subsystems. Systems theory suggests that existence is defined in relations, and that the human psyche, as a system, defines and "re-cognizes" itself through relations. The sense of meaninglessness can be analyzed in terms of lost or altered relations. The experience called "personality crisis" is understandable in terms of rapid change in internal relations, external relations, or both. In such cases, the information mapping through which a psyche "re-cognizes" itself has not adjusted to alterations in some of its subsystems or to new stresses from external relations.

Freedom

These considerations are related to the often debated problem of freedom of choice. The problem is more manageable within a general systems perspective. Instead of "will," let us consider control systems. The human organismic system has many levels of interrelated control subsystems. One of these subsystems has a unique capacity for becoming aware of itself and monitoring some of its own functions. It also has the capacity for symbolic representation of a variety of possibilities, both actual and potential. At first glance, it might appear that this control subsystem is free to choose among these possibilities. However, the real problem emerges when one asks on what basis the choice is to be made. The answer is on the basis of the programmed guidelines established for such selection. Were this all, one could speak of a freedom to function, but only as determined. Faced with a choice for which there are no mapped patterns of selection, individuals seek ways of mapping the new situation and test possible relations. If they succeed, the information map will include a new pattern to determine selection. When conflicting demands arise from various internal subsystems or external relations, the degree of stress usually determines the selection. So far, we have described the human psychic system simply as a very adaptive morphogenic (form-creating) control subsystem that gives the human organism a high level of functional autonomy. However, the fact that it is an ordered, organized, process system means that processes, including behavior patterns and decisions, are systemically determined, even when novel adjustments are involved. But, that, too, is not the whole story.

The most intriguing questions arise in connection with the capacity of the human psychic system for what we now call modeling and simulation. The human psychic system has evolved its capacities for abstract symbolic representation to such a high degree that, with the aid of electronic subsystems, it can play an endless number of simulation games. When these games are played in the context of what we call pure mathematics, pure physics, or any other pursuit at that level, there is considerable freedom to make a variety of choices for all kinds of reasons. The choices are free because the consequences are not related to the survival of the human organismic systems playing the games. This example suggests that freedom is not something an individual has or does not have; it is something that can be developed by degrees.

However, what happens to this kind of freedom when the results of the modeling and simulation arrive at an optimal conclusion and it is time to make an existential choice? Is the psychic system then trapped once more into a systemic set of control systems that will accept or reject on the basis of its mapped information? Certainly this is most often the case, but there remains still another consideration.

If one's level of self-awareness includes an understanding of how to modify the mapping of its own control system, and if the conclusions of abstract simulation and modeling appear to produce desirable results, then it is possible for a highly aware psychic system to produce the conditions necessary for altering control subsystems to adjust to a new set of selection criteria. Put differently, the individual becomes both programmer and computer in a process-adaptive change. This is not to oversimplify the problem. Systems theory suggests that the failure of much contemporary education and therapy results from the tendency to focus on one subsystem level or another. Behavior modification, existential psychotherapy, and psychoanalysis all tend to treat individuals as if they were one dimensional. We need an educational system and systems of therapy capable of working a problem at the level of neurons, neurological patterns, or symbolic systems. The closer we come to this level of self-awareness, the more we are free to adjust our control systems to new guidelines in relation to goals determined with a high level of detachment. From this perspective, freedom is again seen to be something a human psychic system develops in degrees.

Degrees of Existence

This line of thought leads us to another shift in perspective. Since Aristotle set down the basic laws of logic, they have become an integral part of Western thought. They have been criticized, but they come close to what might well be regarded as common sense. On the other hand, the pursuit of an understanding of the systems of the universe requires descriptions of systems in terms of degrees of differentiation. Perhaps the most radical assertion is that existence is best understood as a matter of degree.

If we accept that existence is understood and defined in terms of relations and that nothing can be known to exist about which no relation is known, then it is easy to show that complex systems come into existence by degrees. As I have just argued, freedom for an individual psychic system develops by degrees. Evidence suggests that psychic systems develop slowly over periods of time, and we do not know what a fully developed psychic system might be. It is still easier to see that social systems evolve through degrees of relations. As a matter of convenience, one must determine some point at which to classify something as a system, but for accurate analysis this is more of a hindrance than a help.

It may sound silly to say that one thing exists more than another, and there is no need to defend this kind of statement. It makes more sense to say that one thing exists in more relations than something else, and it is helpful to say that everything exists in certain relations but not in others. For example, there have been many arguments over when a fetus is alive and when it becomes human; simliar arguments continue over when an individual may be considered dead. Another such discussion arises over whether certain social systems were truly civilizations. In fact, one can find illustrations of such problems on the edge of almost any taxonomic generalization.

The advantage of thinking in terms of degrees of existence is that it leads one to ask questions that avoid needless confusion. If it is asserted that a ghost exists, we should ask in what relations it exists. Clearly, it exists as some sort of symbol in a number of psychic systems, and it may be a powerful force in those relations. If someone tells me that my anxiety is all in my head, I can reply that it also relates to several other parts of my organismic system and is strong in those relations. If one keeps in mind that systems must be defined in terms of internal and external relations, then the requirement to specify the relations in which a system exists can clear away much confusion.

The Mind–Body Problem

The so-called mind–body problem is still regarded as one of the great unsolved problems of philosophy. However, if we take each aspect of the problem and ask for the relations in which it exists, the problem becomes less formidable. It provides no information to say that an individual has a mind and a brain and then to ask how they relate. Psychic systems, which process incoming information in relation to stored information and provide an output of information, are directly related to chemical and electrical processes in the neurons of the brain and nervous system. We are learning that different parts of the brain are directly related to some functions of our psychic systems. We know that chemicals can alter psychic systems and that processes in psychic systems can alter body chemistry. If we did not try to call some of our subsystems "body" and other of our subsystems "mind," there would be no problem to discuss. Systems and subsystems exist in different relations.

Religious Experience

Current social science suggests that all societies around the world have had some kind of religion as an integral part of their culture. The common pattern is for these religions to include beliefs in various kinds of spirits and gods that are different from forms of ordinary existence. Many arguments and books surround the question of the existence of these supernatural beings. From the perspective just outlined, we must ask in what relations their existence can be defined.

Clearly, they exist in relation to the psychic systems of the believers, but that is only to say that they exist as symbols in a symbolic subsystem. There is no publicly verifiable information received through what we would consider verifiable empirical sources. Consequently, our only source of information for analyzing their relations, internal or external, is through the experience of the believers, that is, within their psychic systems. If we attempt to analyze the relations in which these entities exist, we must analyze their relations within the psychic systems for which they exist. In every case, gods and spirits relate as control systems with reference to standards of human relations, constraints on some psychic subsystems and their output, and processes for equilibrating stresses within psychic systems. It is interesting to note that social systems have programming processes designed to ensure the potency of these symbols or gods within the psychic systems of emerging generations. Zeus and Apollo now exist in different relations than they once did, as interesting symbols of historical significance. Gods who enjoy powerful relations in psychosocial systems often lose those powerful relations through the stages of civilizational development. It never has been very useful to argue about the existence of one god or another, but it can be very useful to analyze the relations of the psychic systems for whom they exist.

What seems clear is that psychic and social systems require control systems and processes for equilibrating stress within individual psychic systems. The control systems and processes must have relations wihtin the psychic system that are powerful enough to constrain antithetical variety and to relieve stress. These processes cannot be purely intellectual. They must be able to function in relation to many levels of psychic subsystems. In this connection, it is commonly agreed that such powerful control and process symbols cannot be intentionally created. They are generally produced through some set of circumstances involving powerful stress on a psychosocial system and preserved through the induced stress accompanying the programming and reinforcing of individual psychic systems.

However, several branches of Eastern religion have developed the notion that control and equilibration in psychic systems is best achieved through a level of awareness and an acquired disciplined control of the whole organismic system. Some Western psychotherapy has moved in the same direction. It is impossible to predict what will emerge from the present fragmentation of religion and the transformation of psychosocial systems associated with it. It is interesting to note, however, that the suggestions made here concerning levels of awareness and degrees of freedom have much in common with some newly popular Eastern religious disciplines and some Western psychotherapy.

What does seem clear is that science is beginning to equilibrate the stress caused by its fragmentation. A growing number of books and papers indicate that systems thinking will be the foundation for that integrative process. Where science has ventured, philosophy has usually followed. Whatever satisfies the psychic needs of humanity in the future, an enlightened acceptance of our place in this system of systems and of the responsibility to equilibrate our inner psychic stresses will be at its center. General systems theory, understood as a philosophy of the world and of life, can provide a map in terms of which that acceptance could give the same sense of reunion and belonging that has been the consistent goal of religious experience.

Index

A

Accommodation, psychic, 77
Adaptation, psychic, 76-77
Alienation, 101, 104
Amino acids, 10, 30
Animism, 82
Artificialism, 82
Assimilation, psychic, 77
Assumptions of general systems thinking, 219-221
Atmosphere, 55-56
Atomic orbits, 13
Atoms, 12-19

B

Behavioral psychology, 67
Bonding
  covalent, 19
  electrostatic, 19
  social, 150, 195
Boundaries, 5, 18
  of civilizations, 208

C

Carbon chains, 19-25
Carbon cycle, 50
Carbon dioxide cycle, 50
Cell, 53
Characteristics of systems, 2, 219-221
Civilizations
  boundaries of, 208
  cause of the development of, 207-208
  characteristics of, 190
  development of, 189
  disintegration of, 215
Closed systems, 30
Common mapping, 159-161, 178-180
Comparative civilizations, problems of, 177
Competition, 36
  social, 180
Conditioning
  operant, 69-74, 93-94
  respondent, 68-90
Constrained variety
  in psychic systems, 86
  in social systems, 172
Constraint, 3, 4
Continuum (continua), 120-124
  of development, 24
  of differentiation, 29
  of uniqueness, 30
Control, 38-41, 55
  learning and, 115-118
Covalent bonding, 19
Creativity, 61-62
Crowds, 159
Cultural model, 104-106
Cultural pluralism, psychic organization and, 155-157
Culture lag, 106, 107

D

Demonization, 113, 124, 169, 172-173
  polarities and, 124
Density, 8
Dependence, 47-48
Development, systemic, 192
Developmental continuum, 24
Developmental process in social systems, 204

Developmental stages
  in psychic systems, 77–80
  in social systems, 197–204
Deviation, see Constrained variety
Differentiation, 29, 41
  continuum of, 29
  in psychic systems, 153–158
  lack of, in perception, 81
  social, 153–158
Discrimination, 53–55, 69
Disintegration, 107
  civilizational, 124
Dissociation, 82

E

Electromagnetic radiation, 2, 11
Electrostatic bonding, 19
Empirical method, 106, 107–108
Energy, 2–4
  net, 37
Entropy, 10
Environment, psychosocial, 117
Equifinality, 39–40
Equilibration, 4, 17, 27, 36–38, 114
  in psychic systems, 83, 125
  social, 201, 206
Equilibrium, 4
Escape conditioning, 71
Essential self, 133
Evaluation, 96–97, see also Differentiation, Discrimination, Exclusivity
Exclusivity, 7, 53–55
Existence, 26, 227
External relations, 6
Extinction, 69

F

Feedback, 36–37
Fragmentation of knowledge, 214
Freedom, meaning of, 226

G

General systems model, 219, 221
General systems theory, 1, 109
  assumptions of, 219–221
General systems thinking as philosophy, 211
Generalizations, 69, 216
Gravity, 27
Groups, transitory or recurrent, 159

H

Habituation, 68
Hierarchy, 8
Hinduism, 102–103
Hydrosphere, 50

I

Identification, 134–140
Identity, 134–140
  cultural, 152
  loss of, 138–139
  personal, 133–140, 144, 165, 170
  and relations, 130–131
Independence, 47–48
Inertial energy, 27
Input, 6
Integration
  of knowledge, 151–152
  of psychic systems, 118–120, 158–161
  social, 158
Interdisciplinary education, 111
Interface, 6
Internal relations, 6
Invariant functions, 76–77
Isotopes, 17

K

Knowledge
  fragmentation of, 214
  integration of, 151–152

L

Learning, 60, 63–67, 95–97, 168
  control subsystems and, 115–118
Left brain, 97–98
Levels of systems, 52
Loss of identity, 138–139

M

Mapping, 43
  common, 159–161, 178–186
  psychic, 87
  social, 155–156
Mass, 3
Maturation, psychosocial, 84
Meaning, 105–106
  personal, 143–145
Methodology, psychological, 111–113
Mind–body problem, 228
Model, cultural, 104–106
Molecular systems, 19–25
Molecules, 19–25
Morphogenic systems, 40
  psychic, 127
Morphostatic systems, 40
Motivation, 118
Mythopoeic world views, 212–214

N

Necessary sequence of psychic stages, 77
Negative reinforcement, 71

Index

Negentropy, 10
Net energy, 37
Nitrogen cycle, 51
Novelty, 7, 99
Nuclear energy, 27
Nucleic acids, 20–24, 30

O

Objectivity, 131, 145–146
Open systems, 30
Operant conditioning, 69–74, 93–94
Orbits, atomic, 13
Organization
  psychic, 76–77
  social, 159–160, 162, 171–173
Output, 6
Oxygen cycle, 50

P

Participation, 82
Particles, 3, 4, 12
Periodic table of elements, 14
Personal identity, 133–140, 144, 165, 170
Personal meaning, 143–145
Philosophy, general systems thinking and, 211
Photosynthesis, 34–35, 50
Polarity(ies), 4, 5, 128–130
  cooperation–competition, 52–53
  and demonization, 124
  dependence–independence, 47–48
  equilibrium–stress, 29
  individuality–participation, 5
  morphogenic–entropic, 45–46
  subsystem autonomy–system dominance, 5, 49–51, 140–142, 172
Preoperational stage, 80
Pressure, 8
Primitive societies, 128, 178
Problem creation, 173–174
  specialization and, 174
Problem perception, 161–162
Process, 6, 17, 28
Proteins, 20, 30
Psychic organization, cultural pluralism and, 155–157
Psychic stages, necessary sequence of, 77
Psychic systems, 63–65, 87–88, 109–110, 112
  integration of, 118–120
  organization of, 121–124
  physiological basis of, 92–96
  social systems and, 120–122
  unity of, 137–138
Psychological problems, 87–88

R

Rational world views, 212–214
Rationalism, 106

Recognition, 136–140, 143
Reinforcement, 68
  negative, 71
Relation, 3, 4, 6, 26
  external, 6
  internal, 6
Relational universals, 218
Relativism, 217
Religious experience, 229
Reproduction, 43–44
  social, 167
Respondent conditioning, 68
Right brain, 97–98
RNA, 20–24, 53

S

Selectivity, 53–55
Self, 89–90, 112
Self-conscious awareness, 107–108
Self-consciousness, 139–140, 165
Semipermeable membrane, 53
Sensitization, 68
Sensorimotor stage, 78–80, 93–94
Social bonding, 150–195
Social category, 164
Social competition, 180
Social equilibration, 201, 206
Social mapping, 155–156
Social specialization, 162, 192–194
Social stress, 155, 160–162, 192, 201, 205
Social systems
  primitive, 128, 178
  psychic systems and, 120, 149
  simple macro, 128, 178
  transition of, 188
Social tension, 168, 205
Socialization, 167
Specialization, 41, 162
  social problem creation and, 62, 174, 192–194
Spirit, 89–90
Spontaneous recovery, 69
Stress, 30, 114
  motivation and, 118
  optimum, 187, 188
  social, 155, 160–162, 192, 201, 205
Structures, 6, 25
Subjectivism, 101
Subsystem(s), 8
  autonomy of, 179–180, 183
  system dominance and, 5, 49
  production of, 192
Sugars, 20
Supersystem, 8
System(s), 1, 5, 8, 29, *see also* Social systems
  autonomy of, 42, 49, 181–183, 196–197
  characteristics of, 219–221
  closed, 30
  dominance of, 42–43, 49

System(s) [cont.]
  levels of, 52
  molecular, 19-25
  open, 30
  psychic, 63-65, 87-88, 109-110
  stable state, 11
  steady state, 33
  world social, 210
Systemic development, 192

T
Temperature, 8
Tension, social, 168, 205
Transcendent perspective, 145-147, 155, 170-171
Transmission, 6

U
Uniqueness, 30, 42
  continuum of, 30
Universe, as hierarchy of complexity and novelty, 222-223
  open ended, 223

V
Valence shell, 15
Variety, 3, 7
  constrained, 7
  in psychic systems, 86
  in social systems, 183-188

W
Water cycle, 50